Interpreting the Function of Stone Tools

The quantification and computerisation of microwear analysis

Roger Grace

BAR International Series 474
1989

B.A.R.

5, Centremead, Osney Mead, Oxford OX2 0DQ, England.

BAR 474, 1989 : Interpreting the Function of Stone Tools.

ISBN 0 86054 608 X

Price £16.00 post free throughout the world. Payments made in dollars must be calculated at the current rate of exchange and $8.00 added to cover exchange charges. Cheques should be made payable to B.A.R. and sent to the above address.

For details of all new B.A.R. publications in print please write to the above address. Information on new titles is sent regularly on request, with no obligation to purchase.

Volumes are distributed from the publisher. All B.A.R. prices are inclusive of postage by surface mail anywhere in the world.

Printed in Great Britain

CONTENTS

LIST OF FIGURES

CHAPTER 1

CHAPTER 2

CHAPTER 3

CHAPTER 4:

LIST OF PLATES

The photographs taken at 200 magnifications represent 360 x 240μ of the flint surface, and at 100 magnifications they represent 720 x 480μ.

CHAPTER 1:
HISTORY OF USE-WEAR ANALYSIS

The development of the analysis of the function of stone tools (hereafter referred to as use-wear analysis) began by the use of analogy with ethnographic tools. This involved finding ethnographic tools that resemble, in form, prehistoric tools and assuming the prehistoric tool was used for the same function as that documented for the ethnographic tool. These analogies were primarily based on parallels with American Indian tools because of the similar morphology of prehistoric and Indian tools. The main problem, apart from assuming a direct correlation between form and function, is that many prehistoric tools have no ethnographic analogues, notably hand axes and burins.

The next stage was to compare wear traces between prehistoric and ethnographic tools. Gould, Koster and Sontz (1971) observed that the wear traces on Australian Aboriginal adzes (purpuna), used for planing hard wood, appeared similar to those on Quina scrapers, and therefore inferred that the Quina scrapers were used on hard wood. The wear on the Quina scrapers was not differentiated from manufacturing traces and no hard wood was present in the environment from which the scrapers come, the Wurm glaciation of South West France. So even though ethnographic analogy is largely responsible for tool types being called scrapers, borers etc, no direct form equals function relationship has been demonstrated, or even that similar wear traces on ethnographic tools can be paralleled with the wear on prehistoric tools. Though these functional terms are regarded merely as labels for classification purposes, an example of the misuse of functional names for morphological types is found with scrapers as, typologically, both concave and convex scrapers are classed in the same group of tools, but they could not have been used used in the same way because the two kinds of edges are not mechanically useful for the same task.

Experimentation was used to obtain functional information by first assuming a function for a tool and then testing that hypothesis by using the tool in the manner assumed and assessing its viability and efficiency. For example the Danish experiments in using Neolithic axes for forest clearance. However, only the assumed function was tested and possible alternatives were not, so that the efficiency of the tool for a specific task is tested rather than the functional capability of the tool. It may well be the case that a tool is more efficient for an alternative task than the one tested.

For an example of a small experimental programme to test the efficiency of different types of tools and different edges see Appendix 1.

Edge analysis involves taking the working edge as the unit of study rather than the whole tool. Wilmsen (1968) measured 2,139 artifact edge angles and he claimed that the analysis showed there to be 3 modes: 26-35 degrees, 46-55 degrees and 66-85 degrees. This is an example of the use of quantitative methods to analyse data that produced edge angle modes rather than assuming a function for the tools. The meaning and significance of the apparent modes remain a matter of interpretation. Wilmsen interprets the activities that these edge angle modes represent as follows: 26-35 degree angled edges were used for cutting, 46-55 degree angled edges were used for hide scraping/heavy cutting and the 66-75 degree angled edges were used for wood and bone working.

Another example of edge analysis was the work carried out by White and Thomas (1972). Here they were looking for types, as the New Guinea tools which they were studying did not fit into any established morphological typology. They asked some New Guinea people who remembered making and using stone tools to classify their tools. Their classification was based on function in that the initial differentiation was made between flakes that were usable and those that were not. Then they separated the usable flakes into those suitable for hafting and ones to be hand held. The flakes suitable for hafting were then separated into ones to be used as vegetable shredders and ones used for drilling. The hand held ones had a great deal of variation in their overall morphology and the New Guinea people said that they chose a tool for a particular task according to the presence of a suitable edge on the tool. The morphology of the whole tool only played a marginal part in the decision of which edge to use for a particular task.

The New Guinea people's classification of the tools was submitted to statistical analysis, from which it was suggested that the edge angle of the tools was important in that they tended to choose tools with a particular edge angle for a specific task, though not being cognisant of such a process themselves. White and Thomas then went on to derive a classification of the prehistoric material based on such attributes as edge angle. This procedure constitutes one of first attempts to produce a functional classification, though in this case the intention was to use it to study relationships between assemblages, rather than for functional reconstructions of tasks being carried out on the sites.

Cantwell (1979) separated scrapers from mid-west America as being used on hard wood or hide by the presence of edge wear. Her assumption was that working hard wood would produce more edge damage than working hide. Hard wood was considered the worked material because there were no bone or antler artifacts present, though preservation was good. The measurement of the edge angles of these two groups demonstrated that the wood scrapers had a mean of 61 degrees and the hide scrapers a mean of 70 degrees. Thus implying a correlation between edge wear and the angle of edges used for the same activity.

After the the translation of Semenov's Prehistoric Technology in 1964 the emphasis in use-wear analysis centered on the use of microscopy for studying the effects of use on the edges of tools. At first this was mainly concerned with low power (i.e. <100 magnifications) looking principally at edge wear. Broadbent (1979) with his colleague Knutsson used microscopy in conjunction with looking at edges using attributes such as edge angle, profile, and "placement of edges in respect of tool length and width" (ibid,81). They studied the tools in order to discover particular attribute sets based on experiments and then separated the tools into those used for scraping or cutting. This work was done in relation to quartz scrapers from Lundfors, Sweden. The experimental programme consisted of *mechanical* experiments where they attempted to control such variables as contact angle (by using pliers as hafts), standardising stroke length and pressure, and the duration of use was measured as number of strokes. Low power microscopy (up to 80x) was used mainly to ascertain how the edge angle changed during use. They noted that edges tended to stabilise, so that weak (i.e. acute angled edges) or irregular edges quickly became rounded and edge angles stabilised after use. Tools having edge angles <55 degrees that were used on wood stabilised at 70-80 degrees. Tools having original angles of >80 degrees sustained little edge wear. Scraping hides with tools having angles >70 degrees "barely functioned" (ibid,84) as they were not sharp enough to cut through the tissues. Lower edge angled tools would break too easily so that edges of 50-60 degrees were ideal. They suggested that edges were designed by retouching them to the optimum angle, or chosen from debitage as being suitable for a particular task.

They separated edges into modes suitable for use on:

1) soft pliable materials eg. hide.

2) hard heterogeneous materials eg. wood.

3) very hard homogeneous materials eg. bone.

They suggested that, rather than there being distinctive polishes according to worked material, edge wear represents a continuum so that use on different materials can appear the same. For example, a tool heavily used on hide could produce the same edge wear as a tool slightly used on bone. But they considered that it only takes 8-10 minutes of use on a hard material for a tool to exhibit edge wear that is characteristic of use on a hard material, and they assumed tools would have been utilised for at least that length of time, so that they could distinguish between tools that were used on hard or soft materials.

They considered re-sharpening, in which case the edge wear reflects the last use of the tool. But as re-sharpening almost always increases the edge angle the re-sharpening would produce edges more suitable for use on a hard material, so that a hide scraper would not be re-sharpened but another tool would be made in order to continue hide scraping, the *blunt* one being abandoned. They investigated the edge angles of the tools and considered that the edge angles did cluster at 70-85 degrees, which they interpreted as being used on hard materials, and 55-65 degrees which they considered suitable for use on medium to soft materials (n=553).

They also examined the edge contours (i.e. profiles) of their archaeological sample, putting them into the categories convex, straight, and serrated. Concave was excluded because there were only 9 examples. Convex and straight edges had similar edge angles of 70-80 degrees, whereas serrated (and notched) edges tended to have lower angles, 60-70 degrees. The convex edges had more *hard material* wear than the straight ones.

In the paper by Tringham et al. (1974) experiments were carried out to test the following hypothesis: "A tool made of a specific raw material, whose edge is activated in a specific direction across a specific worked material will develop a distinctive pattern of edge damage of a kind that is recognisable on the edges of prehistoric tools" (ibid,178). This was tested by mechanical experiments where it was attempted to control such variables as direction of use, pressure and contact angle. The duration of use was determined by a set numbers of strokes. The problem with mechanical experiments is that mentioned in the last part of their hypothesis; do these mechanical experiments produce edge wear; that "is recognisable on the edges of prehistoric tools", especially when, "The efficiency of such edges in performing a given task was not tested" (ibid,178). So that if they were using edges totally inappropriate to the task (i.e. inefficient), such edge wear which might result would not occur on prehistoric tools and would be irrelevant to the problem of identifying edge wear on archaeological tools.

They attempted to deal with the problem of separating retouch from use-wear by looking for patterning in edge damage. "The distinction of flaking which results from deliberate retouch from that caused by wear from usage has been the source of much confusion. In our opinion the distinction should be made on the basis first of size and second of patterning of the scars" (ibid,181).

The evidence for the patterning of scars was produced by their mechanical experiments, in which they considered the effects of such morphological aspects as edge profiles (ibid,180). But this kind of information is not relevant to prehistoric tools, as such different profiled edges would not have been used for the same task since they are not equally efficient.

Odell (Odell and Odell-Vereeken 1981) is the only practitioner of the low power approach to do a blind test and the results were such that he accepts that the technique is limited to identifying the hardness of the material rather than specific materials. (For a discussion of Odell's blind test see Section 10 of Chapter 4).

A major aspect of low power microwear analysis is the classification of fractures. Kamminga (1982) devised 6 types of fractures in his classification scheme: bending fractures, feather fractures, hinge fractures, retroflexed hinge fractures, step fractures and clefts (ibid,6). He describes various sources for the origin of these fracture variations: hardness of material, yielding or resistant effect of material, angle of tool to material (i.e. contact angle), edge angle of the tool, direction of use and the raw material of the tools (flint, quartzite etc). He mentions an earlier comment of his that "The variations of these fractures might be so complex and unpredictable that fracture patterns would be elusive and not task specific". He then claims that this statement was "unduly pessimistic", in that the condition of activities varies sufficiently for their use-wear to be separated. He compares fleshing of a fresh hide with the adzing of dense wood. His experimental research demonstrated that for the efficient use of tools certain variables must be controlled, for example the edge angle and profile of the tools. This indicates that tool selection by prehistoric people (or design by retouch as mentioned by Broadbent 1979) would have been very important and in fact necessary in order to carry out the task, that is, selecting a suitable edge for the job in hand.

Kamminga looks at fractures in terms of overall size (e.g. large/ small/ micro) and the depth of fractures. More detailed quantification of use-wear fractures has been attempted by other workers (Tomenchuk 1983, Akoshima 1987). But to measure each individual fracture by depth, length, width and classify them into one of the 6 types as used by Kamminga would be as time consuming as using high power microscopy. Without this quantification the fracture classification would be subjective and would vary between analysts, especially as they often use different classification schemes.

Kamminga admits that the problem in classifying fractures is that there is no radical difference between the fracture types (Kamminga 1982,5). They are defined by continuous variables because fractures are a continuum in terms of size and form, and the different categories of fractures can merge into each other so that rather than discrete groups, a continuum is being separated into arbitrary groups. He states that some fractures deny classification because they have features of more than one type (e.g. merging step and hinge termination fractures). However a hinge termination is a discrete attribute produced by the mechanical process of its manufacture and its presence defines a step fracture (see Figure 1).

Figure 1

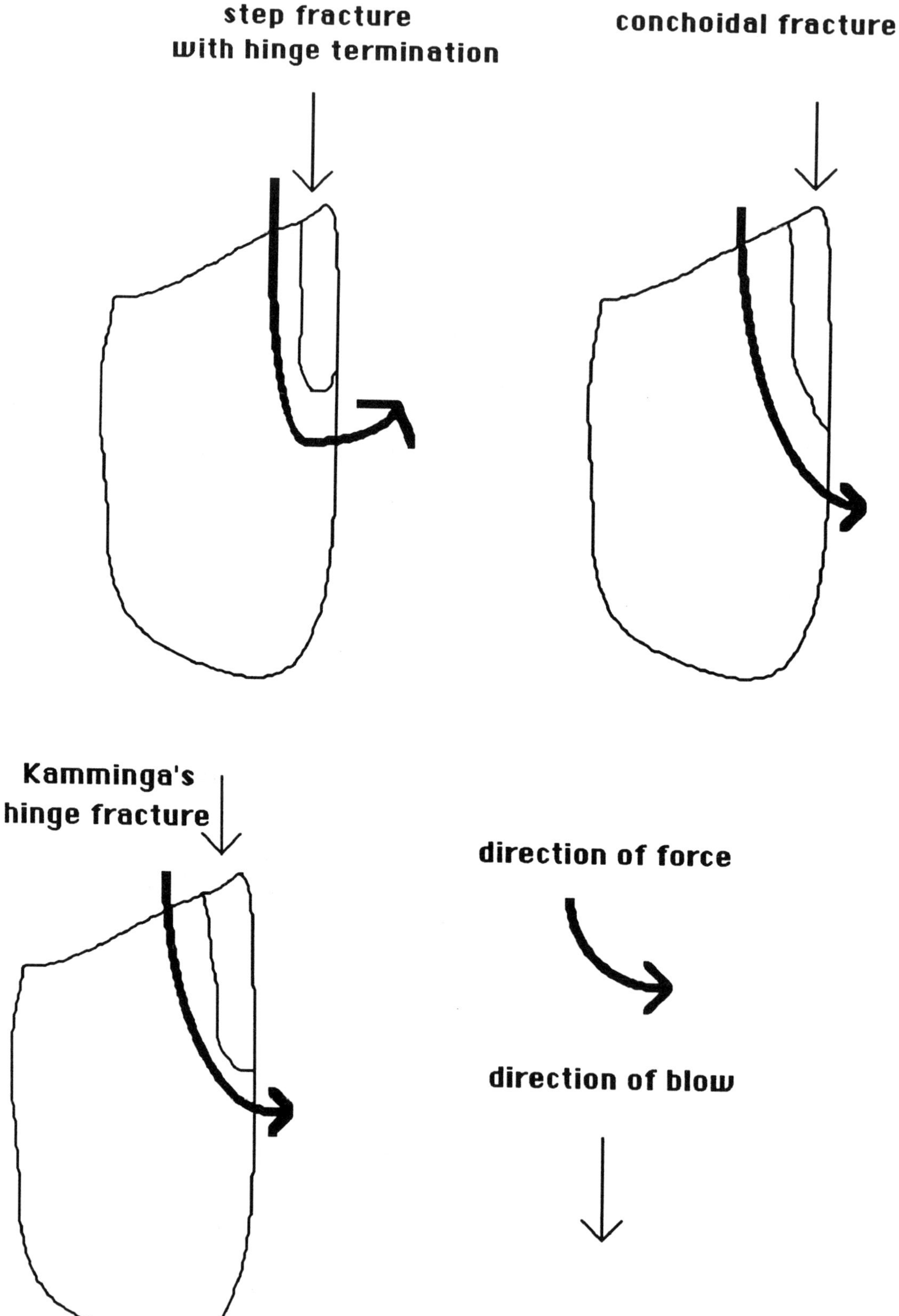

step fracture
with hinge termination
conchoidal fracture
Kamminga's
hinge fracture
direction of force
direction of blow

A hinge termination is created when the force of the blow turns outwards producing a distinct mechanical feature, defined by Crabtree (1972,93) as having an abrupt right angled break at the point of truncation. Kamminga says "I seriously doubt that anyone identifying step fractures adheres to this stringent definition" (Kamminga 1982) and describes a hinge fracture as a sort of smoother step fracture which he says represents a continuum with feather fractures. So there are feather fractures that, as they begin to end more abruptly, become hinge fractures and then step fractures that have a 90 degree termination. The termination angles are not defined and so it becomes a matter of judging the termination angle by eye on very small fractures. This will vary between analysts. The division between conchoidal fracture, where the force of the blow continues through the stone, and step fractures where the force turns outwards to create a second fracture plane, is based on the fracture mechanics of flint rather than any morphological difference between the resulting flake scars. Therefore, the difference is mechanical and the two types represent distinct entities rather than parts of a continuum. The problem with the morphological classification of fracture scars is illustrated by considering step and scale flaking as in Quina retouch. When are these flakes feather flakes with hinge terminations, and when are they step fractures which always have a hinge termination?

High power microwear analysis was developed by Keeley from Semenov's work (1964). High power microscopy involves using magnifications of 100 plus, normally characterising use-wear traces at 200 magnifications, but occasionally using up to 400 magnifications. The extra information gained through the use of higher magnifications centers on polishes. Polish is defined here as the visible alteration of the flint surface so that the reflectivity of the flint surface is increased when viewed through the microscope. Keeley carried out a series of experiments using various tools and he claims to have recognised that specific materials produce distinctive polishes, so that we have bone polish, wood polish, hide polish, etc. The evidence for these distinctive polishes is presented in the descriptions of certain polish characteristics and illustrated with photographs (Keeley 1980).

Keeley originally stated that these polishes are distinctive when certain variables are controlled, in particular the raw material of the tools, so that a prerequisite of any microwear study is an experimental programme using *simulation* experiments with similar stone, preferably from the same source as that of the archaeological material. (*Simulation* experiments attempt to simulate activities assumed to have been carried out in prehistory by using suitable tools in the most efficient manner, rather than mechanical experiments where certain conditions are controlled and which become investigations into fracture mechanics.) If the same raw material is not available, then the raw material should be at least of the same type and grain size. However, since then some microwear analysts have gone further and suggested that such an experimental programme is not necessary, but that the polishes are so distinctive that information from any experimental programme can be used. For example Vaughan (1985), though agreeing with Keeley, uses his experimental reference collection made of Greek flint and of mesolithic morphological types to study flint tools from Cassegros, a Magdalenian rock shelter in southern France.

The main problem with high power microwear analysis is that the descriptions of the *distinctive* polishes are subjective and largely unusable by independent workers. The following chapter describes an attempt to quantify microwear polishes and to test the assertion that the polishes are distinctive and attributable to a specific worked material.

CHAPTER 2:
THE QUANTIFICATION OF MICROWEAR POLISHES

When this research project began high power microwear analysis was based on the identification of polishes that were thought to be distinctive according to the material that the tool had been used to work. So the analysts talked of bone polish, wood polish etc. However, the state of the art of microwear analysis had major disadvantages. The first was the production of comparable results by different analysts. This problem derives from the subjective nature of polish descriptions. Such description terms as *greasy* and with *innumerable pits* (Keeley 1980,43) are often found in the literature. These terms are not specific enough to allow strict comparisons of results produced by different analysts. A *greasy* polish can be bright, moderately bright or very bright (Keeley 1980, Brink 1978) and the term *with innumerable pits* used without any mathematical quantification is a relative term depending on such features as the perception of the analyst, the range of magnification and so on. The relative brightness of a polish is also dependent on such things as the type of microscope in use, any filters that may be employed, and differences in lighting conditions. The Olympus Vanox, for example, as used at the Institute of Archaeology, has two separate adjustments affecting the illumination of the image through the eye piece. The lamp has a variable dimmer control, and the light path through the microscope is also adjustable to allow different levels of light to fall on the subject. Therefore, even if two analysts are using the same type of microscope the viewing conditions have to be the same if the relative brightness of a polish is to be directly compared. The same polish may appear very bright under one set of conditions, moderately bright or even dull under another.

The relative nature of polish descriptions not only made comparative studies difficult, but also extends the process of acquiring the ability to recognise and confidently identify polishes, the necessary *apprenticeship* of the microwear analyst. Anyone beginning a microwear analysis had first to carry out experiments to produce polishes of known origin which they could visually recognise and use as models. In fact, descriptions of micro-polishes only become intelligible when the apprentice has learned his trade. Initially the descriptions are more confusing than helpful and only after looking at a variety of polishes do terms such as looking like "melted snow "(Keeley 1980,56) and features that have "comet shaped pits" (Vaughan 1981,132) have any meaning. The only way of communicating objective information on polishes was through photography. Apart from the technical difficulties of taking good quality photographs, such as the narrow depth of field that decreases as the magnification increases, without any standardisation of magnification and viewing conditions photographs produced by one analyst are often irrelevant to another. Photographs taken at up to 50 magnifications contain little useful information for an analyst working at 200 plus magnifications. The poor quality of reproduction of polish photographs in many journals makes them inadequate for conveying sufficient information.

This means that rather than being able to build on work already carried out, every microwear analyst had to start from the beginning. Even when the apprentice became a master of his craft, the accuracy of the functional interpretations was subject to individual ability and experience. A microwear analysis is a repeatable experiment but given the limited resources and the amount of time required it is unlikely that any results will be checked by independently repeating the original analysis.

The quantification of microwear polishes would facilitate their classification by mathematical criteria, rather than by individual expertise, and any microwear analysis so classified would be comparable with any other. This quantification has been carried out with the use of image processing techniques. Extensive research has been done on image processing to facilitate the identification and classification of images. These techniques have been applied to such diverse problems as the classification of cloud formations, lunar landscape features and the improvement of X-ray photographs for medical diagnosis (Darling and Joseph 1968, Hall *et al.* 1971). The identification and classification of sandstone textures has been carried out with an accuracy of 89% (Haralick, Shanmugan and Dinstein 1973). In all these applications the problem is related to that of classifying microwear polishes. The unit of study is an image made up of tone and texture. In this context texture is represented by the distribution of tone in the image because in a two-dimensional image we visually perceive texture as the spatial relationship of different tones. Where there is little tonal variation in an image the texture is perceived as smooth, where tonal variation between discrete features is marked, a coarser texture is indicated.

The first stage in using image processing to classify microwear polishes is to convert the image into a format suitable for computer input. This is carried out by a scanning digitiser. The photographic negative is placed on a rotary drum and a narrow beam of light is passed through to a photo-cell which measures the photographic density of each resolution cell, resulting in an array of grey level scores. A development of this research was the use of a frame grabber that allows for the instantaneous digitisation from a video camera eliminating normal photography from the process. This technique is discussed in detail below

Figure 2a represents a feature on a flint surface. This image is digitised, (that is, each resolution cell is assigned a grey level); equivalent to laying a grid over the image and calculating the grey level intensity for each cell, as in Figure 2b. The grey level histogramrepresents the frequencies of the grey levels in the array (Figure 2c).

Figure 2

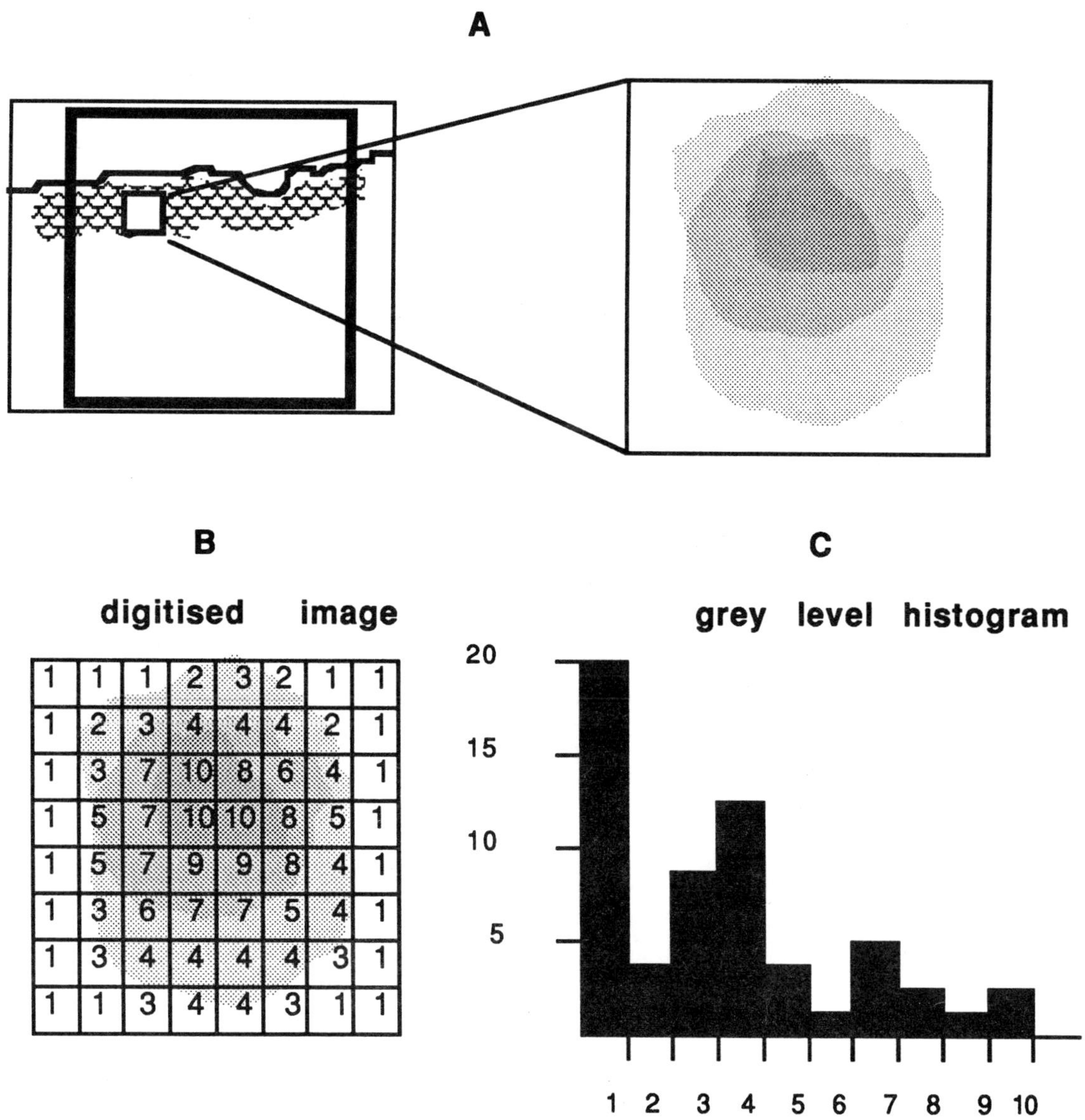
A
B
digitised image
1 1 1 2 3 2 1 1
1 2 3 4 4 4 2 1
1 3 7 10 8 6 4 1
1 5 7 10 10 8 5 1
1 5 7 9 9 8 4 1
1 3 6 7 7 5 4 1
1 3 4 4 4 4 3 1
1 1 3 4 4 3 1 1
C
grey level histogram
20
15
10
5
1 2 3 4 5 6 7 8 9 10

The image of an unused flint surface (Plate 1) can be characterised by its grey level histogram. The histograms of two unused flint surfaces (Figure 3a,b) demonstrate the consistency of the histograms in representing the unused flint surface.

Figure 3

grey level histograms of unused flint

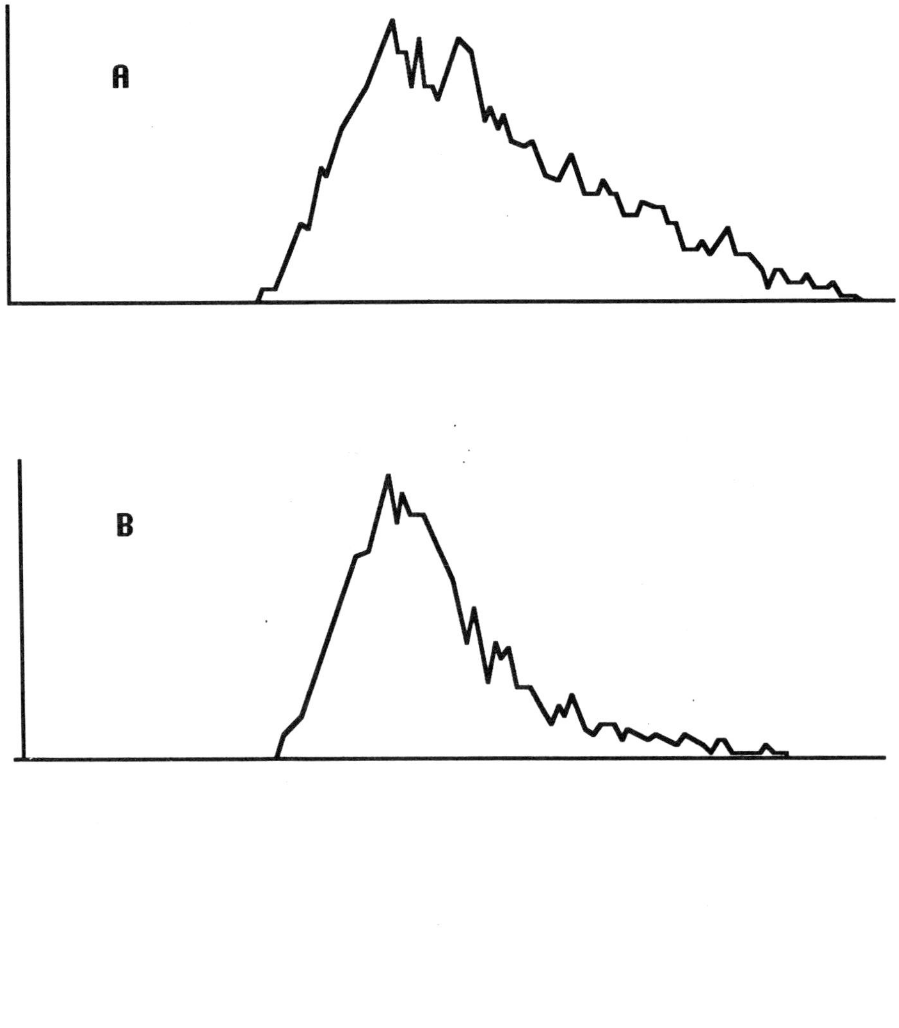

These can be compared with histograms of images of surfaces of flints used on wood (Figure 4), which are significantly different from those representing the unused flint surface.

Figure 4

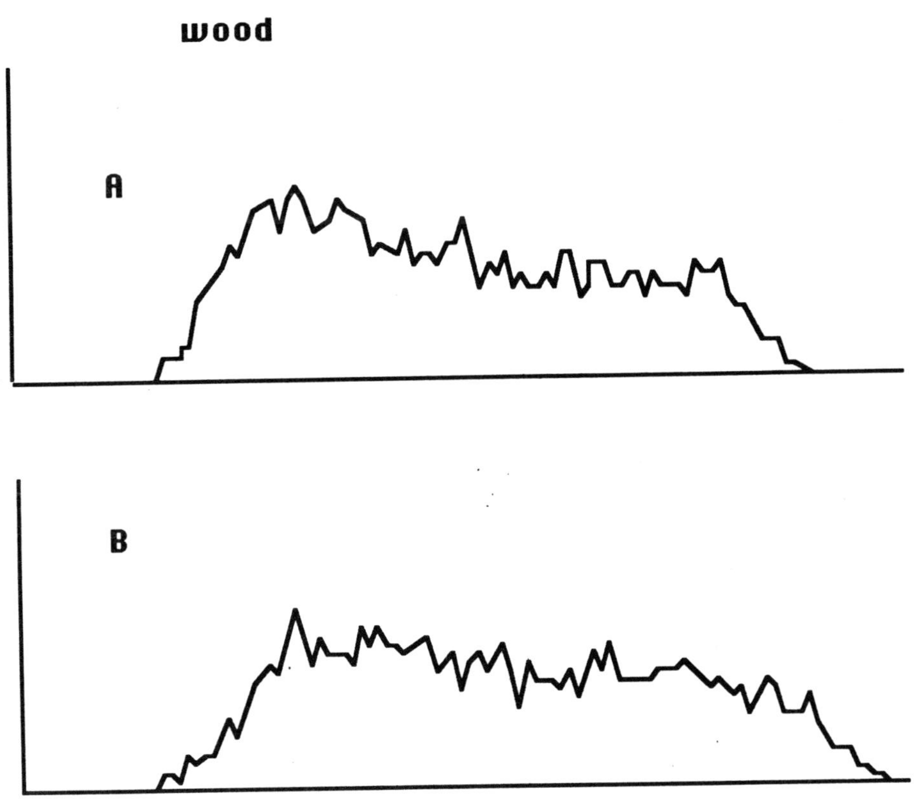

This difference is also present when comparison is made with histograms of images of flint surfaces used on other materials (Figure 5).

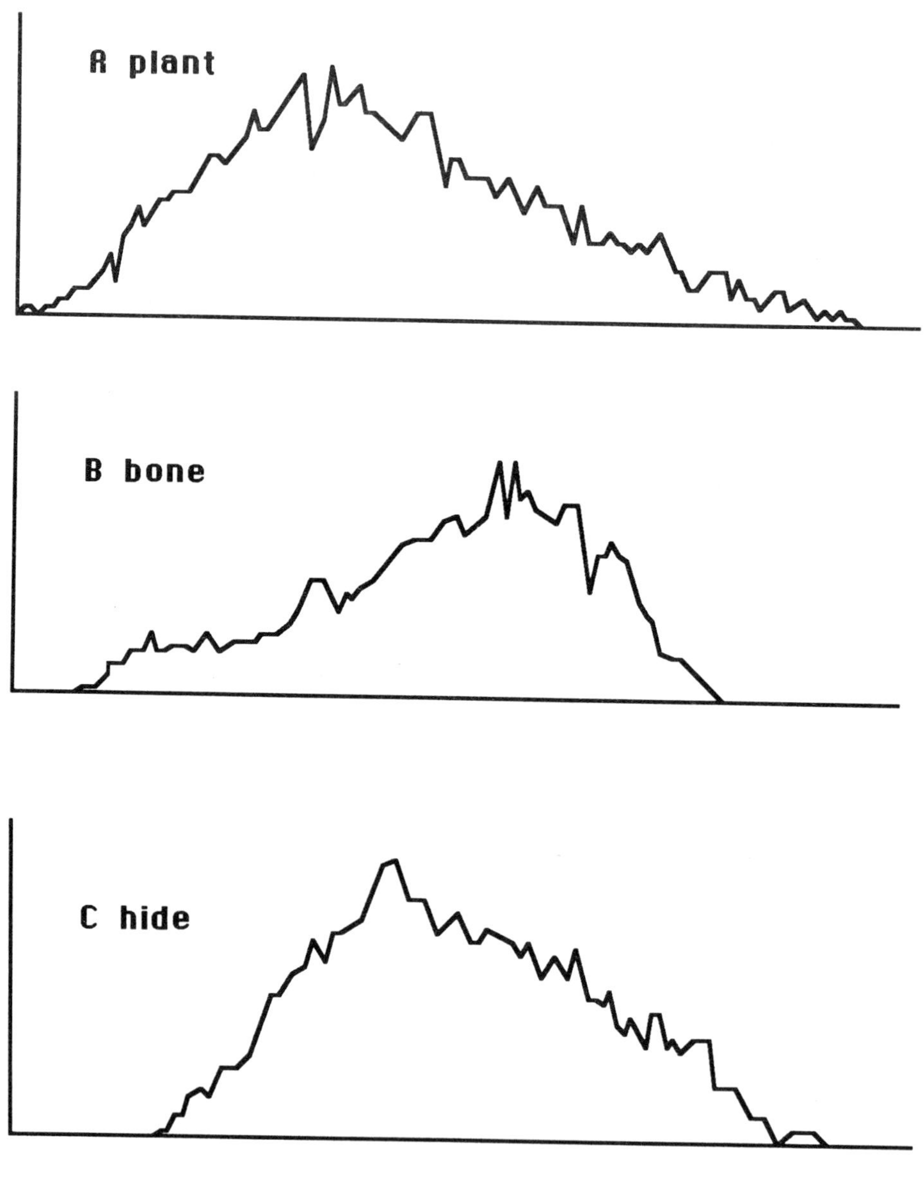

The negative size represents on area of 625x425 µm (Figure 6a), and the central portion of this is digitised over an area of 400x400 µm (Figure 6b). This area is too large as microwear polishes seldom cover such a large area (see Keeley 1980,62). Any size area can be sampled by specifying the number of resolution cells, or pixels. The results presented here are based on samples representing 50x50 µm areas of the flint surface at 200 magnifications (Figure 6).

Figure 6

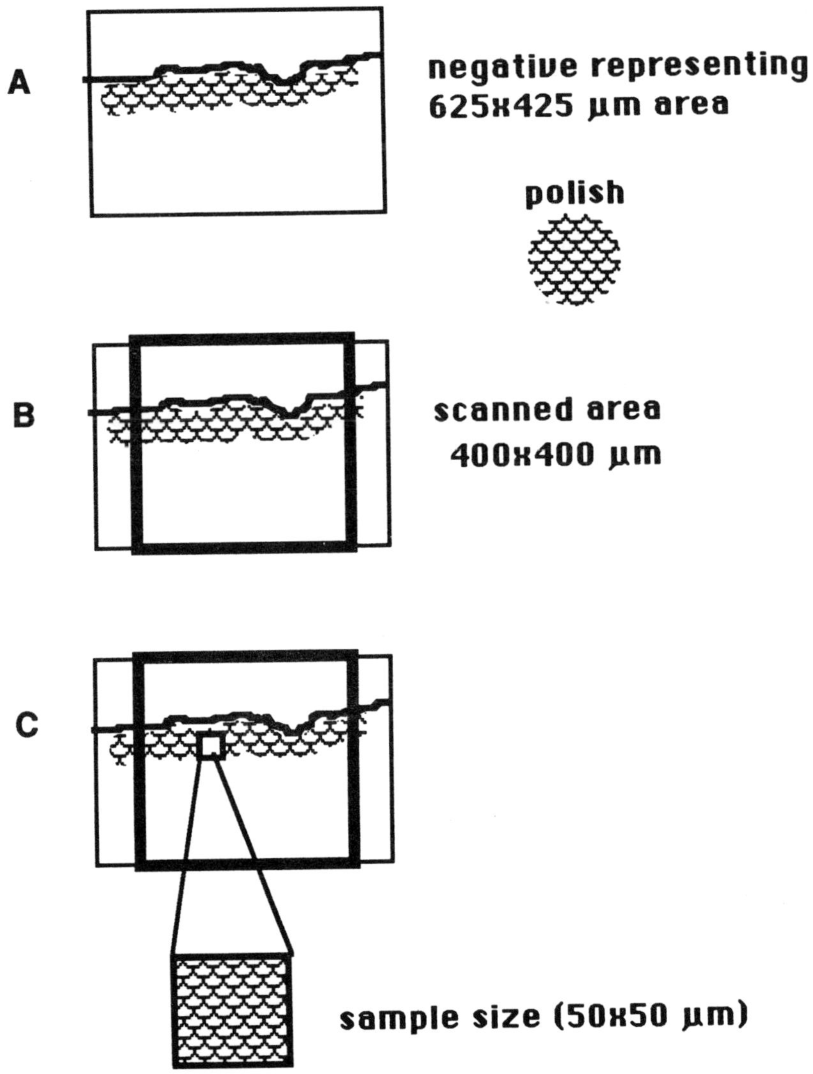

This area is divided into 4,096 pixels and the scanner has a grey tone scale of 0-255, a differentiation that is well beyond the capacity of any human analyst. This means that differences in tone as reflected by photographic density over very small areas can be detected and quantified. The small area of the sample was necessary to obtain samples of completely polished areas as opposed to partially polished areas.

A sample of this size may be subject to local fluctuations. However, the histograms of polish produced by working wood (Figure 4) are from different sections of the same flint surface and their similarity demonstrates that the relatively small sample size consistently characterises the polish. The placement of the sample is specified by row, column coordinates so that any area can be sampled from the 400x400 μm scanned area.

One way in which the distribution of dark and light tones characterises textural features is in the abruptness of tonal transitions. For example, two textural features defined by Vaughan (1981,133) and used as diagnostic features for polish identification are pit depressions, which he describes as holes with sharp edges, and diffuse depressions which he describes as holes with sloping ill-defined sides. In a two dimensional image this textural information is conveyed through changes in tone; pit depressions are dark areas surrounded by the lighter tones of the polish between which the transition is abrupt, while the ill-defined edges of diffuse depressions have a more gradual tonal transition (Figure 7).

Figure 7

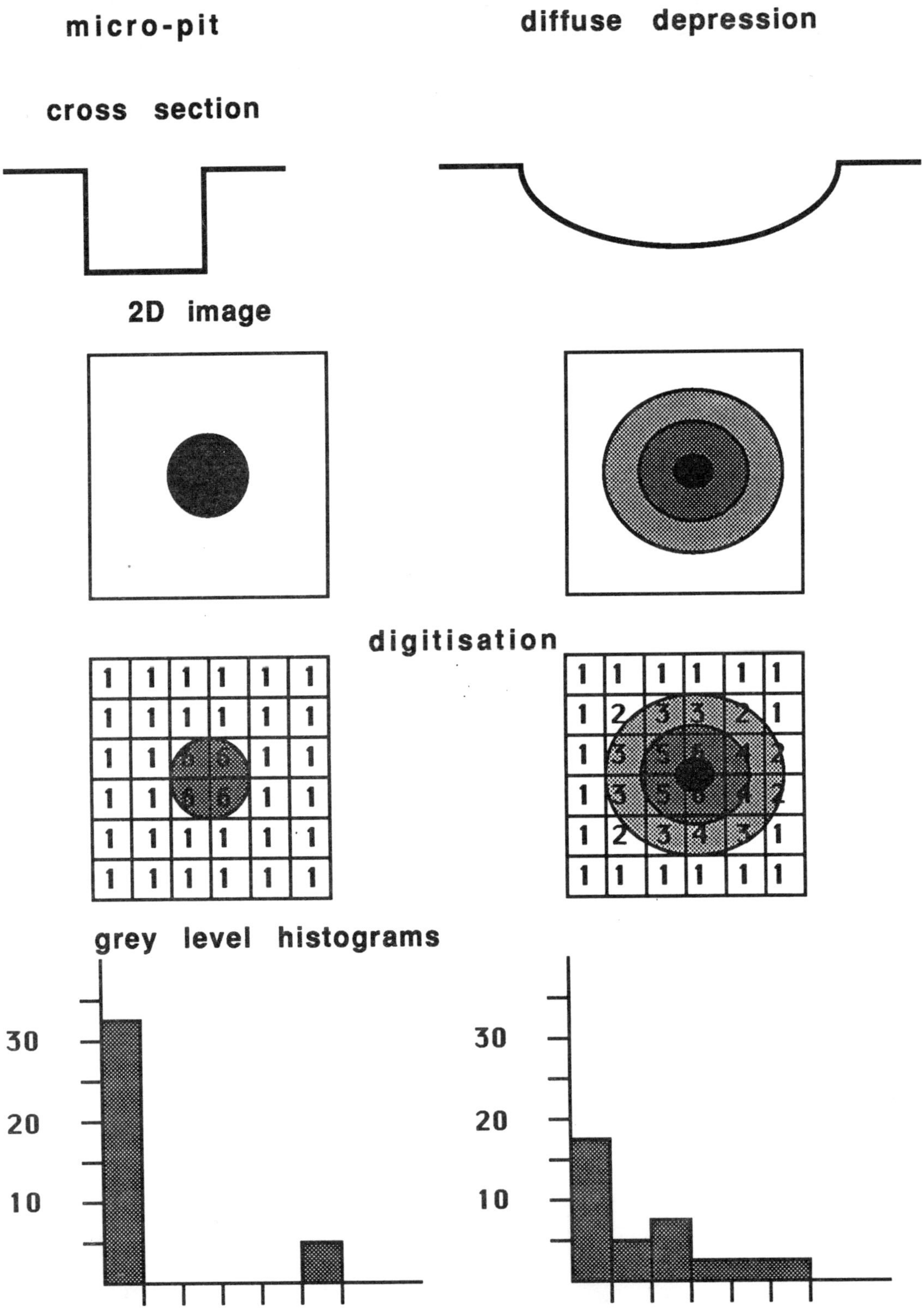
micro-pit
diffuse depression
cross section
2D image
digitisation
1 1 1 1 1 1
1 1 1 1 1 1
1 1 6 6 1 1
1 1 6 6 1 1
1 1 1 1 1 1
1 1 1 1 1 1
grey level histograms
30
20
10
30
20
10

Given that grey level histograms can characterise microwear polishes, a statistic that can be extracted is the difference of a polish from unused flint. This difference is calculated as the square root of the sum of the squared differences between corresponding grey level frequencies divided by the number of grey levels (Figure 8).

Figure 8

$$D^2 = 1/_n \, \Sigma(f_1 - f_2)^2$$

The way in which this distance statistic can be used in classification is by calculating the mean frequencies at each grey level from a number of histograms of the same texture. These mean frequencies can be subtracted from the frequencies of any sample, resulting in a quantification by the **D** statistic of the difference between the texture and the sample, the implication being that the lower the value of the **D** statistic the more probable that the sample polish is of the same type as that for which the means are calculated. Figure 9 represents the mean frequencies at each grey level of 12 images of unused flint, the deviation of these means was very small and the histograms of Figure 3 represent the two extremes.

Figure 9

mean frequencies of unused flint

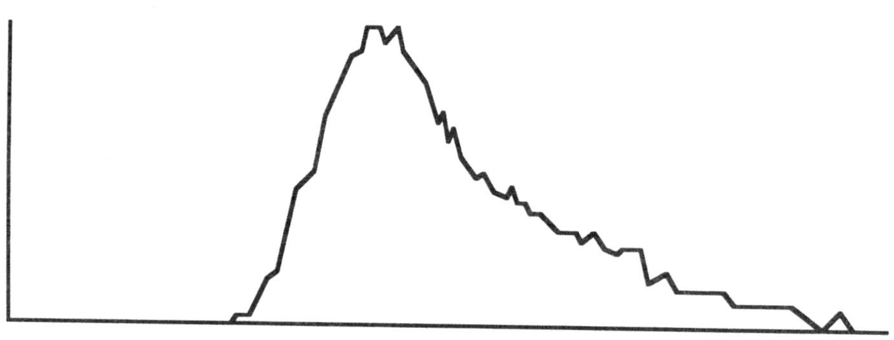

Using this as a base the **D** statistic for a number of polishes can be calculated giving a measure of their difference from unused flint. The values for different polishes are illustrated in Figure 10.

Figure 10

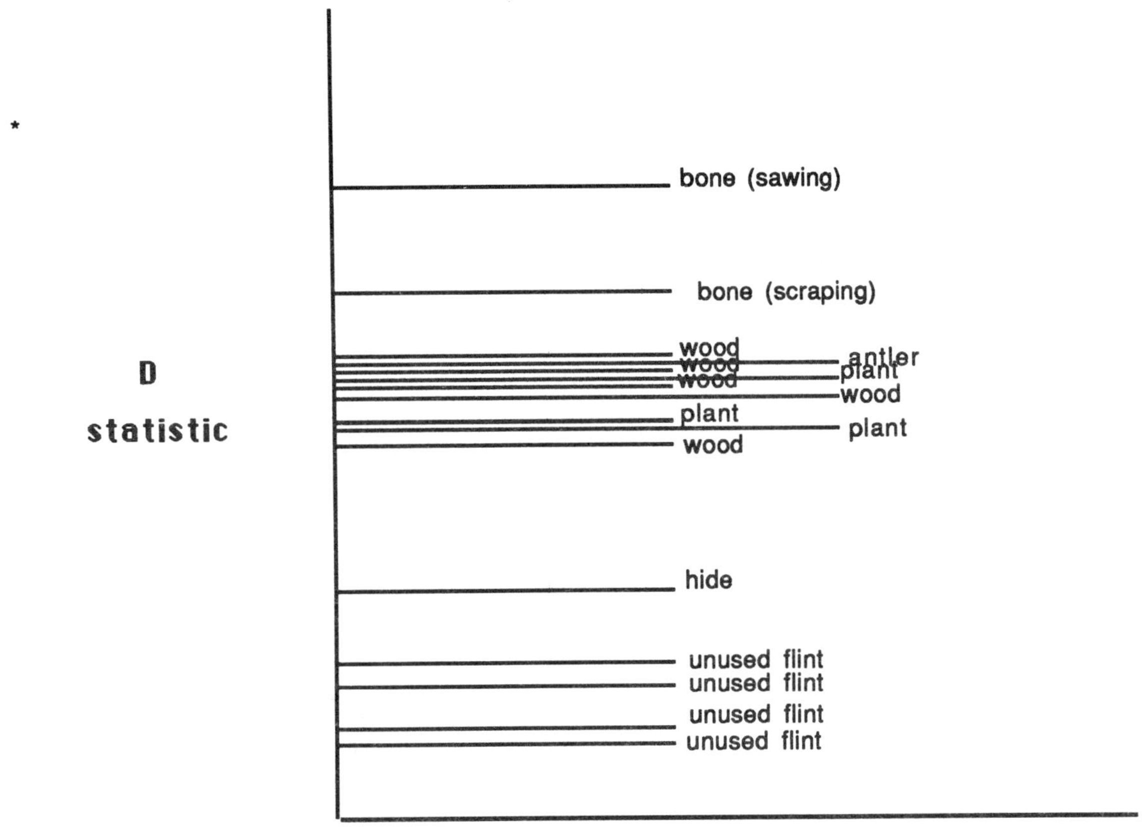

As expected the values for the **D** statistic of samples of unused flint are the lowest, differentiating them from the samples having use wear polish.

A criticism of this method could be that the grey level histograms are unmodified and therefore differences in exposure of the photographs and in processing techniques could be the origin of the differences in the histograms rather than the microwear polishes. The images used in this study have been carefully controlled for consistency in lighting conditions and photographic processing and so the differences in the histograms do reflect polish characteristics. However, when applying the technique to images produced under different conditions, with different film emulsions or by different people, some normalising process would be required. To investigate this problem a selection of histograms were standardised by the use of **z** scores, calculated by subtracting the mean and dividing by the standard deviation at each grey level in the histogram. The distributions have the same mean of **0** and **unit** variance, so that any differences in overall brightness caused by different lighting conditions are eliminated. In Figure 11 the frequencies have been grouped to allow direct comparison to be made and illustrate that differences inherent in the unmodified histograms remain.

Figure 11

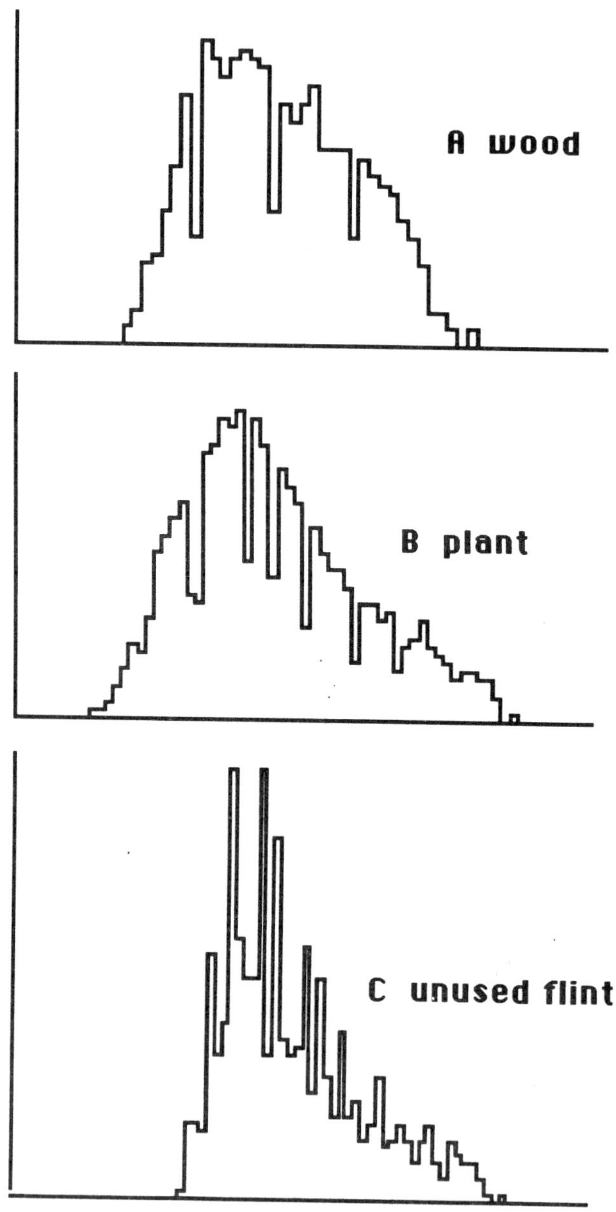

Other statistics can be calculated for texture analysis based on the spatial configuration of the grey levels, rather than their frequencies as for the **D** statistic. To carry out this procedure it is necessary to calculate co-occurrence matrices.

First the grey level histogram is equalised by reducing the number of grey levels so that they have as near as possible the same frequency. For the image in Figure 10, grey level 1 remains the same, grey levels 2 and 3 are combined as group 2, grey levels 4,

5 and 6 are combined as group 3, and grey levels 7,8,9 and 10 are combined as group 4. The array is then transformed using the equalised histogram (Figure 12b). For example, original grey level 3 becomes 2 and original grey level 7 becomes 4 in the transformed array.

Figure 12

A original histogram **B equalised histogram**

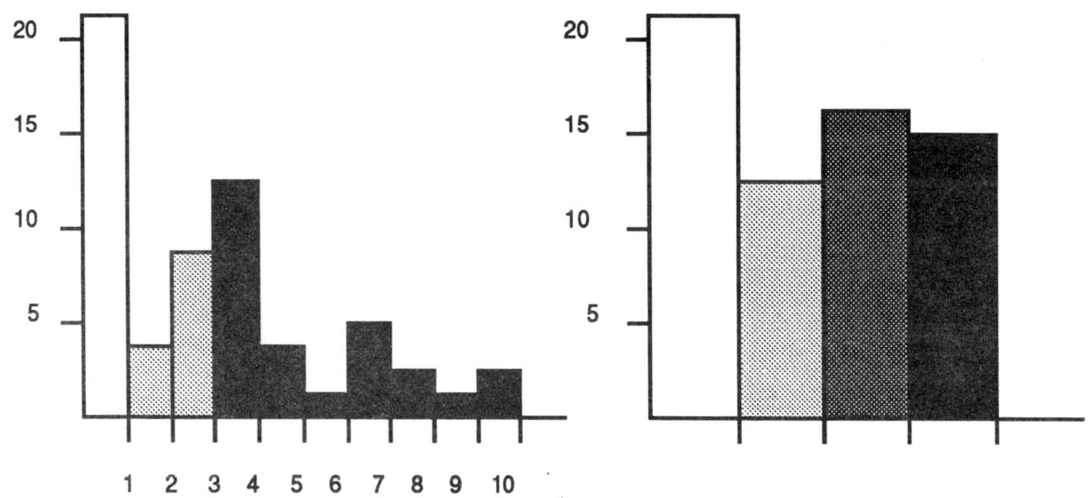

C transformed array

1	1	1	2	2	2	1	1
1	2	2	3	3	3	2	1
1	2	4	4	4	4	3	1
1	3	4	4	4	4	3	1
1	3	4	4	4	4	3	1
1	2	4	4	4	3	3	1
1	2	3	3	3	3	2	1
1	1	2	3	3	2	2	1

This process has the function of standardising the arrays and it also enhances the images by increasing the contrast as represented in Figure 13.

Figure 13

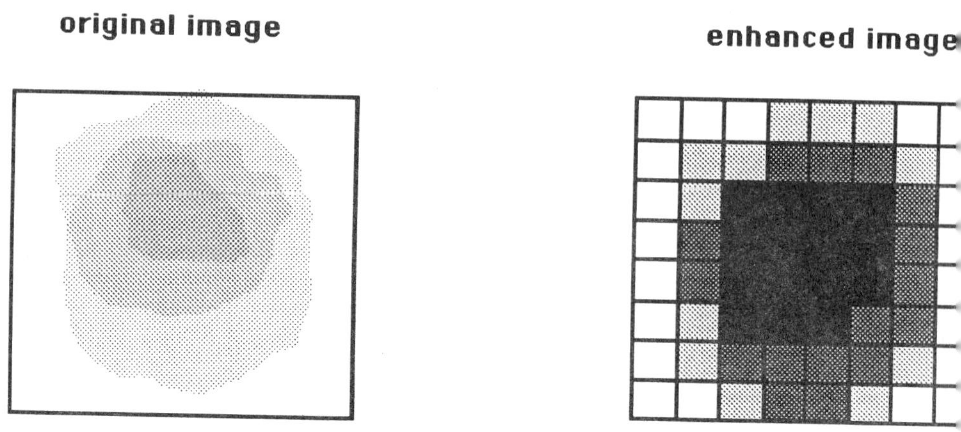

Figure 14 illustrates a similar process for a schematic image of unused flint.

Figure 14

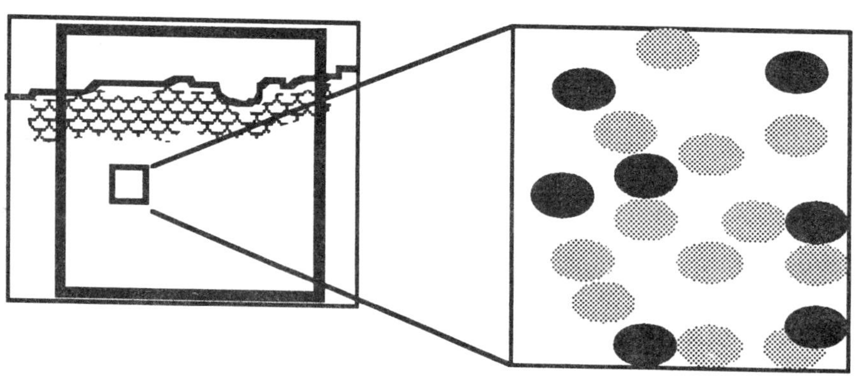

1	1	1	2	1	1	1	1
1	1	1	1	1	1	1	1
1	1	2	1	2	1	2	1
1	1	1	1	1	1	1	1
1	1	1	2	1	2	1	1
1	2	1	1	2	1	1	2
1	1	2	1	1	1	1	1
1	1	1	1	2	1	2	1

A co-occurrence matrix is calculated from the transformed array, as in Figure 15. A vector, that is a distance and direction parameter, is assigned and the combinations, or co-occurrences of grey levels of cells in the array at that vector are calculated and entered in the matrix. In the example (Figure 15) vector 1,-1 is used, that is one cell down, one cell across. From cell **x** this gives cell **y** and the co-occurrence is 4,3. Therefore a score is entered in the 4,3 cell of the matrix. This process is repeated for each cell in the array.

Figure 15

calculation of co-occurrences at vector 1,-1

*

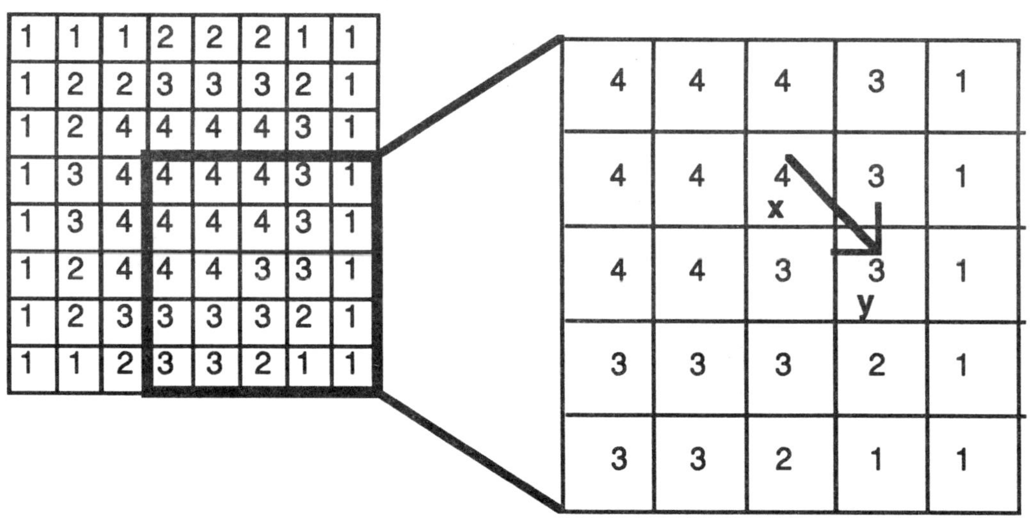

co-occurrence matrix

	1	2	3	4
1				
2				
3				1
4				

Figure 16 illustrates 3 examples of co-occurrence matrices calculated for vector 1,-1. In can clearly be seen that the co-occurrence matrices reflect the spatial information contained within the images. Naturally these are extremely simplified examples for the sake of clarity and for readers who are not statisticians or mathematically minded (for a detailed explanation of this technique see Pratt 1978).

Figure 16

images | co-occurrence matrices

A

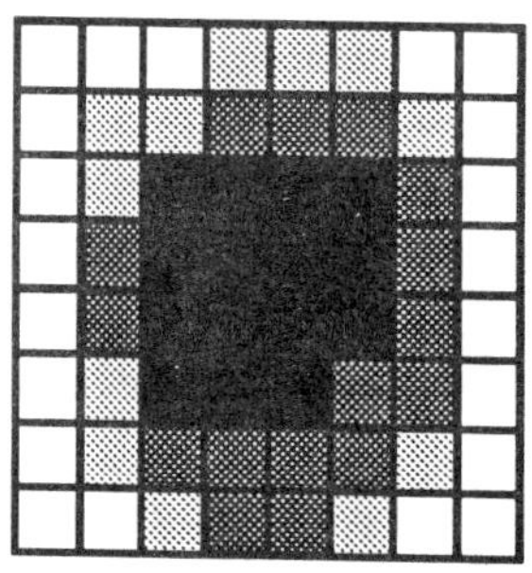

	1	2	3	4
1	2	5	3	0
2	2	2	2	3
3	5	2	3	4
4	0	0	8	8

B

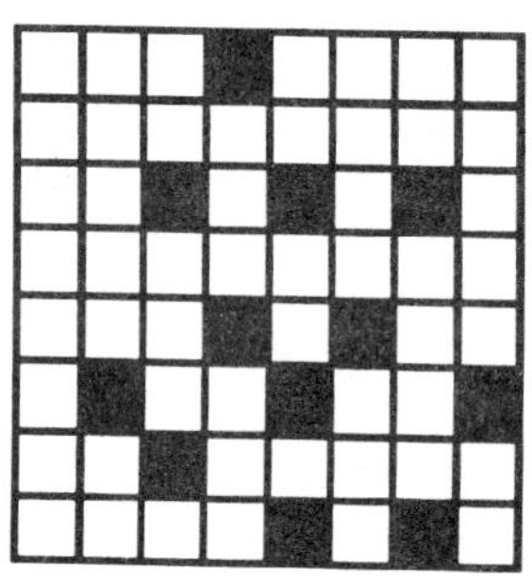

	1	2	3	4
1	31	9	0	0
2	7	2	0	0
3	0	0	0	0
4	0	0	0	0

C

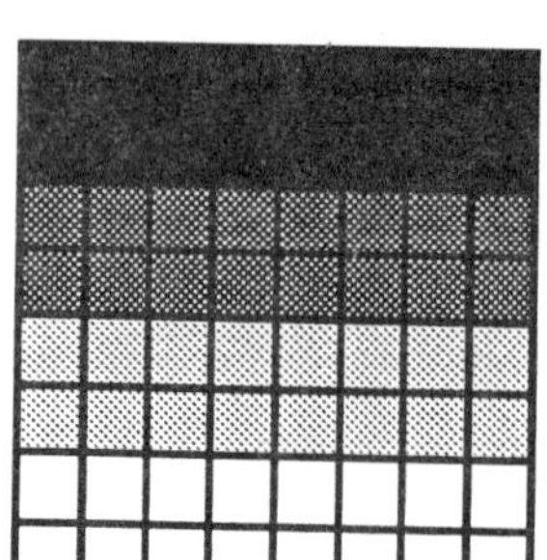

	1	2	3	4
1	7	7	0	0
2	0	7	7	0
3	0	0	7	7
4	0	0	0	7

From these matrices, statistics can be extracted that measure the amount of texture present in the image. For example, if all the cells in the array had the same grey level, all the scores in the matrix would be along the diagonal, as all the co-occurrences would be of the same grey levels, representing a uniform smooth texture. The **CON** statistic is a measure of how the high scores concentrate along the diagonal, giving a value of **0** for the case of a uniform image. The angular second moment **(ASM)** texture feature is a measure of the homogeneity of the image as it is calculated from the spread of high co-occurrence values throughout the array. Thus when the texture is evenly spread it will have many entries of low magnitude as opposed to a less homogeneous texture which will have few entries of high magnitude.

The algorithms used to calculate the statistics are given below.

Figure 17

$$CON = \sqrt{\Sigma_i \Sigma_j \{i-j * p(i,j)\}^2}$$

$$ASM = \sqrt{\Sigma_i \Sigma_j \{p(i,j)\}^2}$$

i = row, j = column, p(i,j) = value in cell.

Therefore to calculate the CON statistic the column number is subtracted from the row number, the result being multiplied by the value in that cell and then squared. The Con value is then the square root of the sum of the values for each cell. The ASM statistic is calculated by squaring the values in each cell and then finding the square root of the sum of those values of all the cells in the co-occurrence matrix. Figure 18 illustrates the values of the CON and ASM statistics for each cell for image A, B and C.

Figure 18

A

CON

	1	2	3	4
1	0	25	36	0
2	4	0	4	36
3	100	4	0	16
4	0	0	64	0

= 17

ASM

	1	2	3	4
1	4	25	9	0
2	4	4	4	9
3	25	4	9	16
4	0	0	64	64

= 15.52

B

CON

	1	2	3	4
1	0	81	0	0
2	49	0	0	0
3	0	0	0	0
4	0	0	0	0

= 11.4

ASM

	1	2	3	4
1	961	81	0	0
2	49	4	0	0
3	0	0	0	0
4	0	0	0	0

= 33.09

C

CON

	1	2	3	4
1	0	49	0	0
2	0	0	49	0
3	0	0	0	49
4	0	0	0	0

= 12.12

ASM

	1	2	3	4
1	49	49	0	0
2	0	49	49	0
3	0	0	49	49
4	0	0	0	49

= 18.52

For example the value for cell 1,2 of the matrix is 1-2 [=-1] *5 (the value in that cell) [=-5] squared =25. For cell 1,3 of the matrix the calculation is 1-3 [=-3] *3 (the value in that cell) [=-9] squared =36. By subtracting the column numbers from the row numbers all values on the diagonal of the matrix (representing co-occurrences of the same grey level) will have a value of 0. The greater the distance of the cell from the diagonal the greater its contribution to the CON statistic, as in the above example so that cell 1,3 has the smaller value (3) than that of cell 1,2 (5) but contributes 36 to the Con statistic while cell 1,2 only contributes 25. The ASM statistic is calculated by simply squaring each value in each cell. Therefore the values for cells 1,2 and 1,3 used in the above example are 25 and 9 as opposed to their Con values of 25 and 36. This means that when there are large numbers of co-occurrences of the same grey level combination the ASM is consequently increased.

Using the above formula for the three co-occurrence matrices in Figure 16 the **CON** values are 17, 11.4 and 12.12; the **ASM** values are 15.52, 33.09 and 18.52 respectively. Therefore the more regular textures of images B and C are similar (11.4 and 12.12) and lower than the **CON** values for the more complex texture of image A. The **ASM** statistics reflect the tonal variation within each image, B having the least tonal variation and consequently the highest value (33.09) Image C has as much tonal variation as image A but is more spatially regular and therefore has the higher value (18.52 as opposed to 15.52).

In this study it was found that the differences between textures of different polishes were most evident at a small scale, and so the **CON** and **ASM** statistics were calculated for a distance of up to 3 resolution cells, which gives 16 vectors and hence 16 values for the **CON** and **ASM** statistics.

Having characterised the micro-polishes, the statistics can be used to classify the polishes by calculating the difference of a polish from unused flint. Samples of textures can be subtracted using the same formula as for the **D** statistic, exchanging the frequency values for the values of the **CON** and **ASM** statistics.

Figure 19

$$Dcon^2 = 1/_n \Sigma(c_1 - c_2)^2$$

$$Dasm^2 = 1/_n \Sigma(a_1 - a_2)^2$$

Dcon and **Dasm** represent the difference of a sample from the model of an unused flint surface. **c** = con values, **a** = asm values

Previously the problem, concerning the **D** statistic, of differences in sensitivity encountered in different film emulsions was discussed. Further research revealed that different batches of film had significant differences in sensitivity that could not be overcome, even by normalising the histograms. Therefore the **D** statistic was abandoned because of these technical difficulties and the results presented are based on the **CON** and **ASM** statistics. For a description of the experimental program and statistical data on which the results are based see **Appendix 1.**

As the value of the **CON** and **ASM** statistics are expressed as differences from unused flint, a scatter diagram drawn using the **CON** and **ASM** statistics as the axes, has the property that the nearer a point is to the origin of the scatter diagram the more similar it is to unused flint. Figure 17 clearly shows how this technique can isolate unused flint textures, as all the samples of unused flint cluster around the origin. It is also clear that polishes produced by contact with the same material do not cluster together.

<u>**Figure 20**</u>

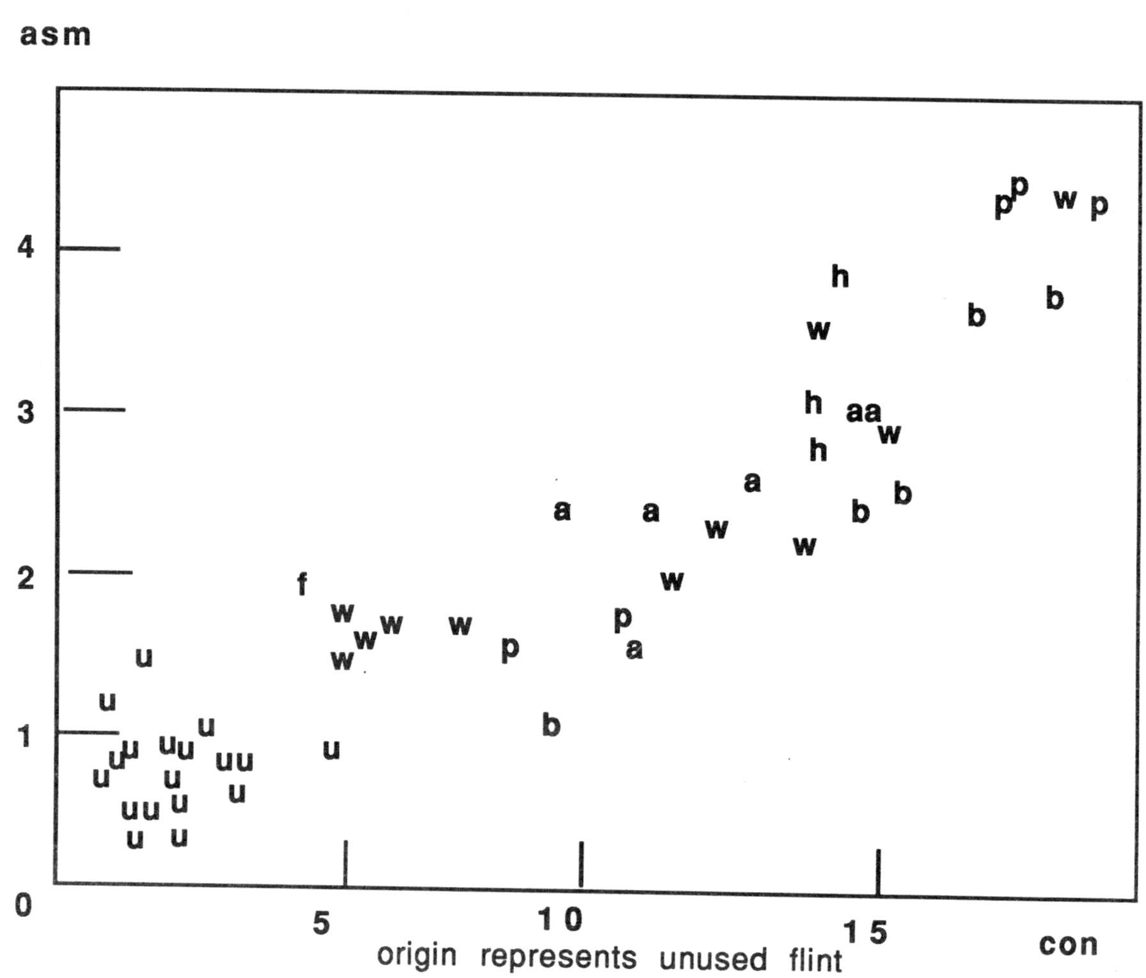

u unused   w wood   a antler   f fish   b bone   h hide   p plant

If polishes produced from contact with the same material are distinctive, then difference values from a model of a particular polish should be able to be calculated by the same process as was used for unused flint. Polishes produced by contact with that material should then cluster around the origin as with the unused flint samples in Figure 20. For example, if the mean **CON** and **ASM** values of a number of polishes produced by contact with antler are calculated and a scatter diagram prepared in the same way as above, the antler polishes should cluster near the origin. Figure 21 is the scatter diagram of polish differences from just such a *model* antler polish.

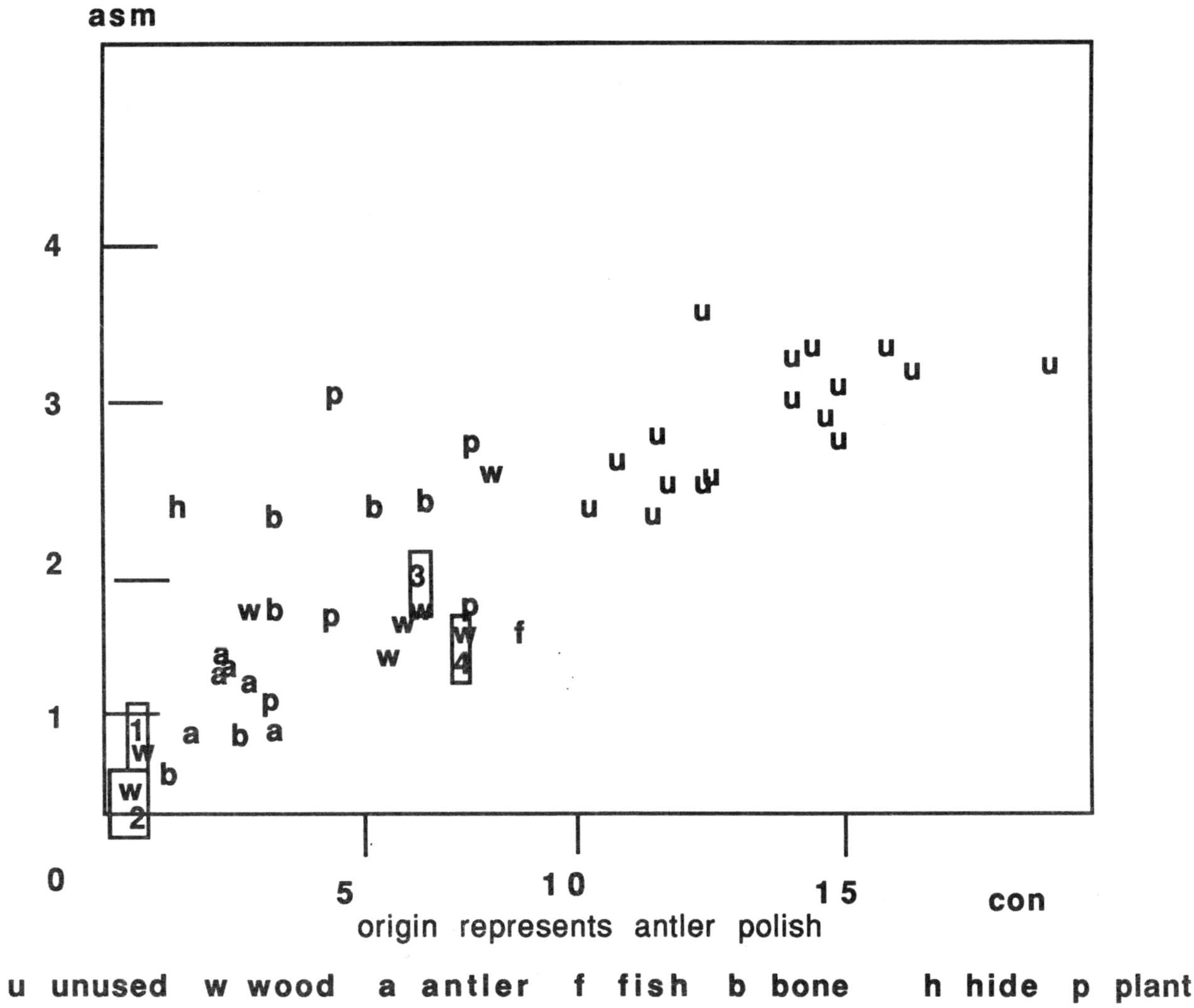

The diagram shows that antler polishes do not clearly separate out, and that some wood polishes cannot be distinguished from antler by these two texture measurements. Indeed the two polishes most similar in this test to the model antler polish are in fact wood. This confirms the general suspicion that some antler and some wood polishes are indistinguishable. This has been reported by several microwear analysts (e.g. Anderson-Gerfaud 1981 v.1,61, Vaughan 1981,144, Moss 1983,87) while Keeley states that "wood and antler polishes can often be quite similar to each other" (Keeley and Newcomer 1977,55). In blind tests wood, antler and bone polishes have also on occasion been confused with one another (Keeley 1980, Gendel and Pirnay 1982). It is also interesting to note that the two wood polishes most like the model antler polish (**1** and **2** in Figure 21) were produced by 20 and 30 minutes use as opposed to the antler polishes produced from between 3 and 12 minutes use. The wood polishes which were clearly separated from antler were on tools used for between 5 and 15 minutes (Figure 21; **3** and **4**). This indicates that tools used on different materials for differing lengths of time may produce a similar polish, and so cannot be distinguished either by optical microwear analysis or by the texture analysis techniques used here.

The intention of applying texture analysis to microwear polishes was to develop an objective method of identifying tool function from the distinctiveness of polishes according to contact material. However, the results described above suggest that the overlap of polish textures produced from contact with different materials is greater than was thought. The polishes are not simply a product of contact with a specific material and other variables seem to be equally important in determining the quantity and characteristics of the polish that is visible to the microwear analyst. The period of time for which the tool was used is an important factor in distinguishing between polishes, and unfortunately in the case of archaeological specimens that information is by nature irrecoverable. The texture analysis results also indicate that other variables, such as edge morphology in relation to use motion, have an effect on polish development. For example, whittling wood with an unretouched flake produces a different texture to that produced by scraping wood with an endscraper. In Figure 21 points **1** and **2** represent flakes used for whittling wood and point **3** represents an endscraper used for scraping wood; point **4** represents wood scraping with a burin.

To check these results produced by texture analysis, that seemed to indicate that polishes were not distinctive according to worked material, further analysis was conducted in conjunction with a blind test (Newcomer et al. 1986).

Ten tools were made and used by an experimenter (M.H. Newcomer) and then handed over to the analysts for identification of the used area, the motion of the tool and the worked material. Five analysts took part in this test and their results are presented in Figure 22.

Figure 22

		USED AREA	MOTION	WORKED MATERIAL
1	Endscraper on wood	4(1 partial)	3 (2 partial)	1
2	Drill drilling shell	1	1	0
3	Piercer grooving antler	5	0	0 (2 partial)
4	Endscraper on stone	4 (1 partial)	4 (1 partial)	0
5	Burin scraping bone	4 (1 partial)	0 (1 partial)	0 (3 partial)
6	Backed blade shot into earth	4 (1 partial)	3	1
7	Unused endscraper	0	0	0
8	Backed bladelet cutting meat	3 (1 partial)	3	1 (3 partial)
9	Blade cutting fish	3 (1 partial)	1 (3 partial)	0
10	Burin scraping wood	4	3	0 (1 partial)

In the scoring for this test a point was awarded for a single specifically correct answer. In addition a partial credit was awarded if the analyst had correctly identified the worked material within a multiple answer, i.e. an answer of bone/antler was regarded as partially correct if the correct result was bone.

These results were very poor, only three tools being identified to a specific worked material out of a possible 50 identifications. An independent investigation of the polish characteristics was made using texture analysis.

Figure 23 is a scatter diagram expressing the difference between samples and an unused flint surface with the addition of the blind test tools which are represented on the diagram by their respective numbers. (Tool number 8 representing a bladelet used for cutting meat will be discussed later).

Figure 23

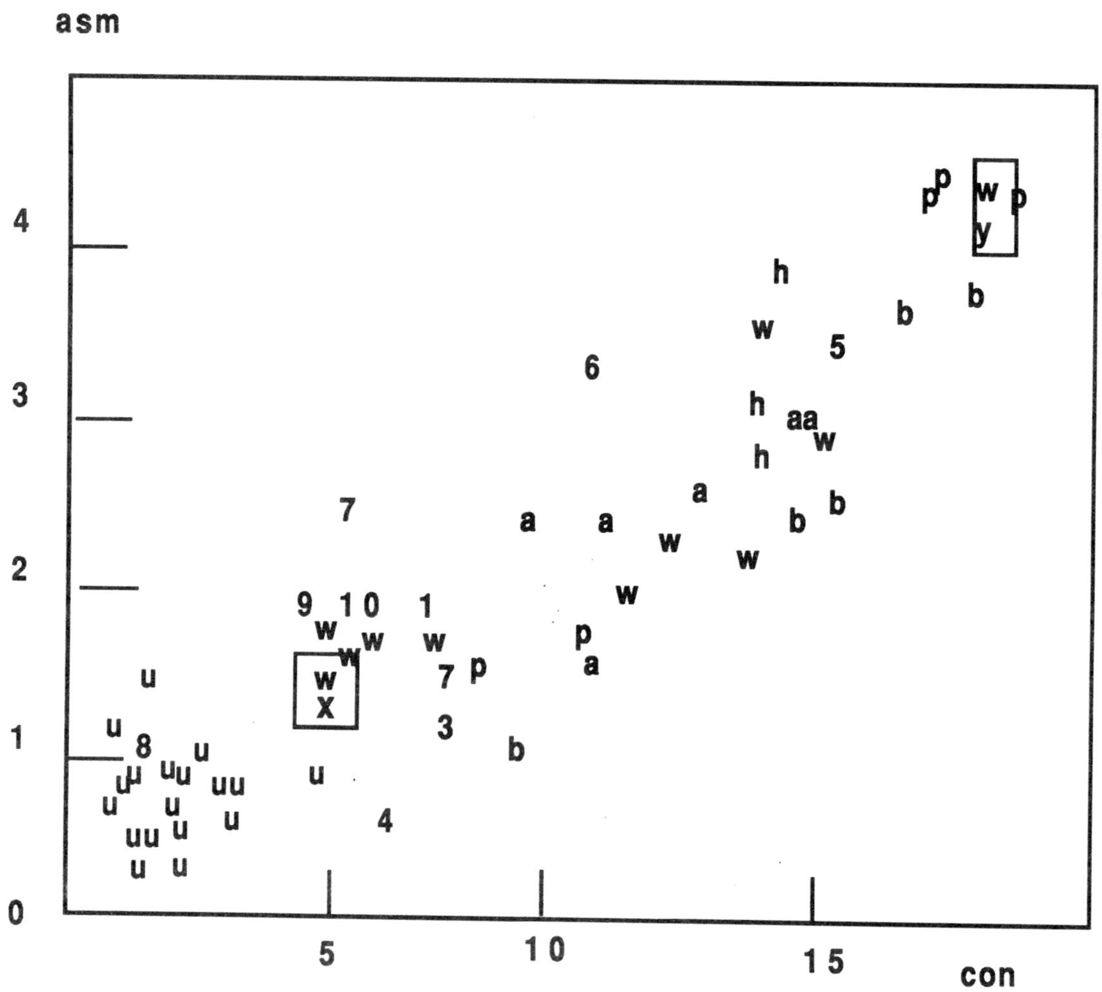

origin represents unused flint

u unused w wood a antler f fish b bone h hide p plant

As with the previous diagrams polishes produced from contact with different materials are spread throughout the diagram, and do not cluster into distinctive groups according to worked material. For example, point **x** is one of the polishes most similar to, and point **y** is the polish most different from, unused flint. But both of these polishes are from the same tool, used for scraping wood for 15 minutes (Plate 2: a,b). Both polishes are present on the mid-point of the used edge, the difference is that point **x** represents polish from the ventral surface and point **y** from the dorsal surface. These polishes are the result of sawing, a use motion that would not be expected to produce different polishes on different surfaces as both were touching the wood throughout the task.

This technique is capable of clustering unused flint textures and distinguishing them from polish textures, even though there is considerable variation between the unused flint textures. Among the samples are materials from four different sources. Included are English chalk flint from Newton and Brandon (Norfolk) and Potters Bar (Hertfordshire), as well as the Seine flint of the test tools.

Using the scatter diagrams (Figures 23 & 24) the position of the test tools can be examined to see how they relate to the results carried out by standard microwear analysis. Figure 24 is a scatter diagram of the difference values of the samples from *antler polish*, with the addition of the blind test tools.

Figure 24

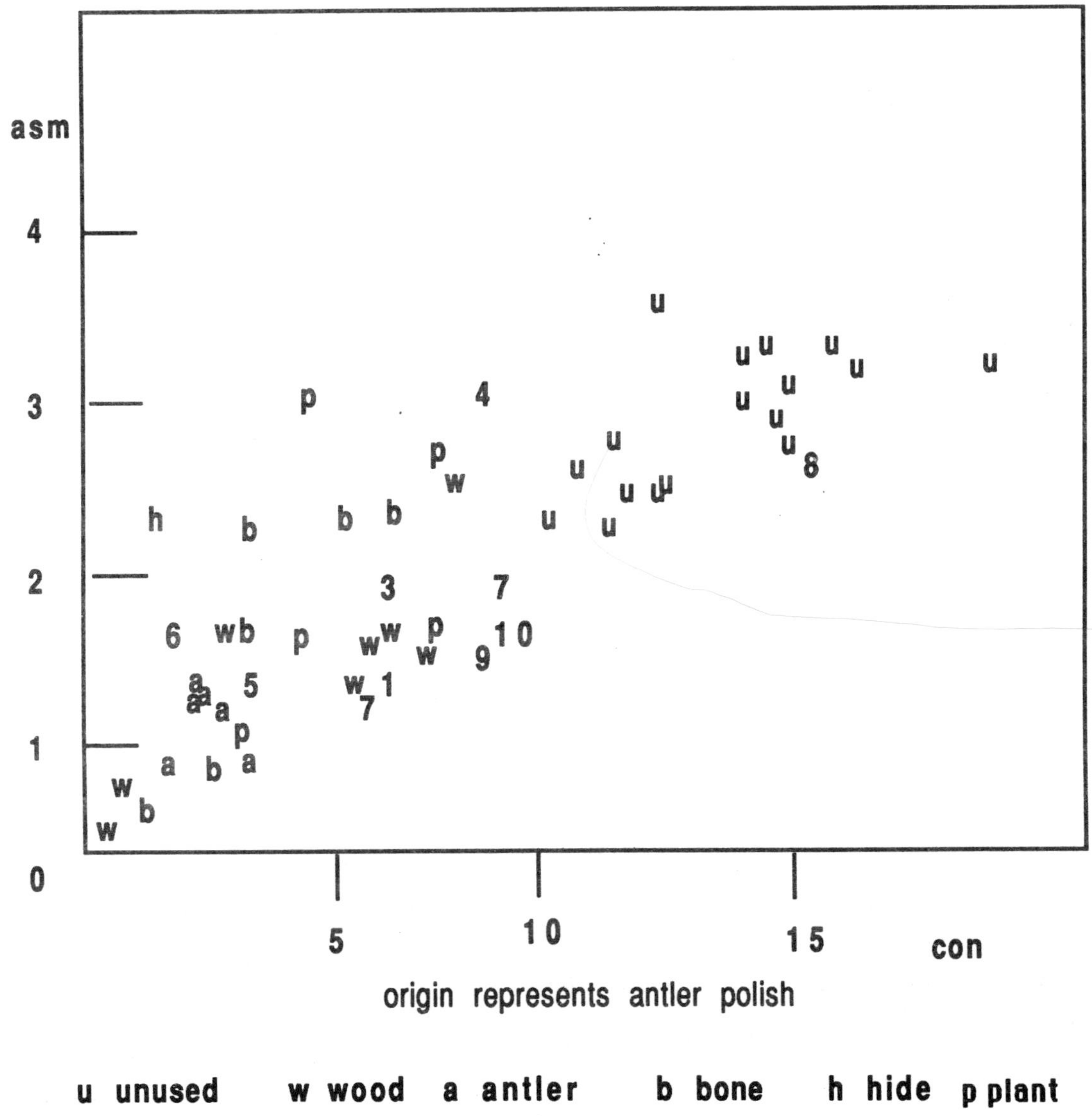

Test tool one was used for scraping wood for 15 minutes and clusters with some other samples of polish produced by wood, even though it visually differs from them.

Test tool two, the drill bit, retained no polish which perhaps never developed or was lost through edge wear; in any case the tool could not be analysed.

Test tool three was used for grooving antler and had only isolated areas of polish. The analysis of one of these small areas places the polish near to the group that contains tool number 1 (used on wood), rather than with other polishes produced by contact with antler.

Test tool four was used for scraping stone for 6 minutes. It does not cluster with other polishes and shows the least difference from unused flint of the test tools having polish. This indicates the lack of polish development which would make it difficult to identify, and no analyst could identify this polish.

Test tool five was used for scraping bone for 11 minutes and clusters with the main group of antler/bone polishes. All the analysts identified this as bone or antler polish, though the nearest point on the diagram represents a polish produced by whittling wood for 20 minutes.

Test tool six was used as a projectile point and has polish similar in texture to that on tool five (used on bone). Although the contact material is unknown, the polish produced is of the same nature and intensity as bone/antler polish, but no further conclusions can be made.

Test tool seven was unused and is the most problematical of the test tools. In most cases the microwear analysts stated that this tool was used and three analysts identified hide polish. The texture analysis found that certain areas of the flint along the retouched edge had a surface that was significantly different from unused flint. An area of the central portion of the ventral surface also had a surface that was unlike unused flint though no polish was visible through the microscope.

Test tool eight was used for cutting meat for 29 minutes and produced no significant surface alteration and so clustered with unused flint. No characteristic polish was identified by the microwear analysts, but some did interpret the function of the tool correctly. This identification, however, was based on the morphology of the backed bladelet, and the absence of a developed polish indicating a worked material which produces little or no polish, a material such as meat (Plate 3: a,b). The correct identification of this tool could not have been based on polish characteristics as the texture analysis appears to demonstrate that no significant surface alteration took place during use.

Test tool nine was used for cutting and scraping fish for 20 minutes, and clusters with wood polishes. This means in terms of texture analysis that it has the same nature and intensity of polishes produced by working wood, and in one case was identified as a wood polish in the test.

Test tool ten was used for scraping wood for nine minutes and occurs in the same group as test tools one and nine, associated mainly with wood.

As mentioned earlier, polish textures do not cluster simply according to worked material, and so the test tools would not be expected to cluster with polishes produced by working the same materials. The significance of their positions on the graph is that it demonstrates why incorrect identifications of polishes can be made.

For example, if a polish has little difference from unused flint this means that the polish is insufficiently developed to exhibit characteristics that could indicate the worked material. Tool four, used for scraping stone, is such a case (Plate 4). Furthermore, if two polishes produced by working different materials are similar to one another, as in the case of tools nine (fish) and ten (wood), then the probability is that they will be identified optically as the same (Plate 5: a,b). Of course, with prehistoric tools these inaccurate identifications would go unnoticed. Only with experimental tools whose function is known is it possible to detect such errors.

These results of the texture analysis of the test tools are based on the texture of the polishes alone, and therefore the similarities and differences are due to polish features such as brightness, smoothness and the spatial arrangement of texture features within the scanned area. The fact that the scanned area is small, but of very high resolution means that sub-visual texture features are being measured and it could be argued that features such as *comet-shaped pits*, edge rounding and reticular patterning would not be characterised by this method. However, these features are not sufficiently consistent in their association with a specific worked material. They are indicators of a possible function, rather than being characteristic of a specific function.

For example, the reticular patterning often associated with wood polish is caused by the original topography of the flint surface, rather than being a product of contact with one material. A reticular pattern can be created on an uneven flint surface by a variety of materials. Conversely, with a smooth topography the same function will not produce a reticular pattern. Edge rounding is specified as a characteristic of dry hide working (Keeley 1980,50), but again it is possible to produce edge rounding with any material of sufficient hardness, as the rounded edge of tool four, used to scrape limestone, clearly indicates. All the analysts noted that edge rounding was present on this tool and four out of five identified the tool as being used on hide, mainly on this supposedly diagnostic feature, as the polish on the tool was insufficiently developed to be identified. Therefore these kinds of features cannot be used as exclusive indicators of a specified material.

If features of attrition and intra-polish distribution are a function of the combination of the original flint topography and the hardness of the worked material, the problem becomes one of how to differentiate between materials that have a similar hardness. To use an example from the blind test, what is the difference in hardness between fresh wood and fish, as revealed by their relative effects on similar flint surfaces? The difference in hardness would be detectable only if the duration of use were held constant. If, as in the case of the blind test, a tool is used for scraping wood for nine minutes and another used for scraping fish for 20 minutes, then the relationship of the hardness of worked material and tool use duration could combine to produce an equivalent abrasion factor resulting in similar polishes being produced by contact with different materials.

One way in which tools having similar microwear textures can be separated according to worked material is by examining the characteristics of the used edges of tools for their suitability for use on materials of different hardness. When the group of textures that cluster around the origin of Figure 24 are examined, they are separable according to contact material by their edge characteristics. For example, the tools used on bone and antler are burins, a piercer and the tip of a blade, as opposed to the backed blade used on plant material and the concave unretouched edges of flakes that were used on wood. This is not simply a correlation of tool type with function, as within the bone/antler working tools the different types of burins, piercers and blades are represented. The tools are typologically different, but the working edges on these tools have similar characteristics whether they occur on burin bits or tips of blades. A burin bit is simply not efficient for whittling wood, and a concave lateral edge of a flake cannot be used for grooving antler. Therefore the edge characteristics of tools can determine their functional capability, so that the function of tools having similar microwear polishes can be identified by their suitability for a particular task according to their edge characteristics. The importance of these edge characteristics and the morphology of tools is demonstrated by the possibility of determining the function of test tool number eight as meat cutting, even though no polish was present.

Further blind tests were carried out by rubbing flints on different worked materials and also by using a cutting motion for a set period of time on uniform blades of flint. The results of these test further challenged the assumption that polishes are distinctive to a specific worked material (see Newcomer et al. 1986).

The results presented here are based on the techniques described above. The technique has been developed further by the use of a frame grabber. The frame grabber allows for the instantaneous digitisation of images directly from a video camera mounted on the microscope. A *live* image of a flint surface is produced on the monitor which has a square, representing a 50x50 µm area at 200 magnifications, at the centre. This square can be centered over the area to be sampled by maneuvering the flint under the microscope, this allows for a preview of the sampled area. The sampled area is then *grabbed* i.e. digitised and stored in the computer as an array. The mathematical transformations and statistics are calculated instantaneously. This not only allows for more accurate sampling and for many more samples to be taken, but also eliminates conventional photography from the procedure. This removes the technical difficulties mentioned earlier in relation to differences in film emulsion sensitivity. Using the frame grabber the analysis presented here has been repeated and many more samples taken. However, these results are not presented in detail because they duplicate those already presented. The conclusions from the analysis with the frame grabber are the same as these presented. That is, polishes do not cluster according to worked material.

Though the frame grabber has not produced any new results it has enabled images of flint surfaces to be captured and stored. Experiments have been carried out using CD-ROM compact disk technology in order to produce a compact disk with a data base of flint images, together with information of edge morphology, edge wear etc. (see Chapter 4). The information is recorded on a specially developed data base program. This allows for the recall of an image according to search techniques that involve stipulating certain parameters eg. all examples of flints within a certain edge angle range, or having specific features like snap fractures (see chapter 4, section 5.), or any combination of attributes. It is hoped to develop this project to produce a data base of images and data of a large number of experimental flint tools to use as a reference collection for functional analysis.

There have been other attempts to quantify microwear polishes. Keeley attempted to measure the brightness of polishes with the use of a light meter as part of the photographic system attached to the microscope (Keeley 1980). There were two main problems, first the size of the area required in order to take light meter readings was a circle of 280 µm in diameter of the flint surface. Keeley states this area is too large "most microwear polishes, even on extensively used specimens (except corn gloss) seldom cover a large enough area of the micro-surface to fill even the 280 µm diameter area " (Keeley 1980,62). Secondly, the readings of the photographic meter are not sufficiently precise "the use of a smaller standard area cut the variation in light intensity so severely that, given the crudity of the microampere scale on the meter, discrimination was impossible " (Keeley 1980,62). As Cook and Dumont (1987,54) comment "Keeley (1980) shows that light reflection measured in microamperes through the microscope is both impractical and inadequate to the task of measuring polishes."

Dumont attempted to quantify microwear polishes with the use of interferometry (Dumont 1982). This work involved converting the visual image to an interference image. The work was never developed to the stage of attempting to quantify interference images of microwear polishes.

Another technique that has been attempted is profilometry which has been used to quantify the topography of polishes which would give a measure of their smoothness. This technique has been used by a number of workers. In Japan by Akoshima (1981) and in Sweden by Knutsson (pers. comm), both these workers found that the precision of the profilometer was insufficient for it to record the topography accurately. This technique is mainly used to discover cracks in metal, particularly developed for testing for metal fatigue; metal has a relatively smooth surface which when cracked produces a significant irregularity that is easy to detect. The problem with flint surfaces is that they are irregular to begin with and these irregularities are produced by the original topography of the flint as much as by any change produced by use. The results produced by profilometry have not shown that polishes produced by use on a particular material have a consistent profile. The techniques of light meter readings, interferometry or profilometry have not produced results that enable the polishes to be classified according to mathematical criteria.

A potential technique for measuring uneven flint topographies in order to examine the differences brought about by use is the application of fractal geometry (Mandelbrot 1982). Preliminary investigations have been carried out to test the feasibility of applying fractal geometry to flint microwear polishes (Rees, Wilkinson, Orton and Grace 1988).

The results of the texture analysis indicate that the formation of microwear polishes is a complex procedure. In order to interpret the function of stone tools the isolation and quantification of a number of variables and an understanding of the interaction between the variables is required. Concentration on polish characteristics is clearly inadequate if the results of functional analysis are to be made more reliable.

Since this chapter was written an important paper on the use of image processing of microwear polish images has been published (Knutsson et al. 1988). Very similar techniques of texture; analysis were used by the researchers which supplement those discussed here.

The procedures for the analysis in the Knutsson et al. paper were the digital scanning of polish images, the calculation of grey level histograms and co-occurrence matrices in a similar way to that reported here. 16 texture measures were calculated for the images, four of which were selected and used to derive a discriminant function. The difference with this process is in the choice of texture measures. The CON and ASM statistics were chosen as texture measures that best reflect the way way in which microwear analysts perceive polish textures through an optical microscope. That is, they are held to measure such features as smoothness, homogeneity of texture, amount of texture, presence of micro-pits etc. as described above. Also these two texture measures were chosen because they had been used in similar types of analysis with some success, such as that on sandstone textures (Haralick, Shanmugan and Dinstein 1973). This was to attempt to quantify the suggested differences between textures produced by contact with different materials. The different approach in the Knutsson et al. paper is that the texture features chosen were those that best discriminated between polishes produced by use on different materials. That is the design set consisted of tools used on plant, wood, hide, antler, bone that were all used for the same length of time (30 minutes). The texture measures employed in the classification were those that best discriminated between these known groups, but there is no description of how these particular texture measures (difference entropy, information measure A, peak transition probability and information measure B) relate to the perception of flint textures by optical microscopy. Even with this selection of the texture statistics that best discriminated between the groups, a number of examples were assigned to the wrong group. For example an unused tool was classified as having been used on hide and a tool used on antler was classified as unused. More examples of incorrect classifications occurred after the classification was Jacknifed (Knutsson et al. 1988, 262; Figure 3. The specific procedures followed are not given). The misclassification of unused flint textures is particularly disturbing as the one distinction that all microwear analysts can agree on is between an unused flint texture and a polished one. The texture analysis presented here shows that using the CON and ASM statistics reliably and consistently distinguishes between unused and used flints (with the one exception of test tool 7) and therefore is perhaps more accurately assimilating the optical perception of the microwear analyst.

The plot of the 32 cases of the design set in the canonical variates plane shows that only textures produced by working bone are distinguished from the remaining textures (ibid,263; Figure 4). Three groups are suggested by the authors, those of 1:bone, 2: wood and vegetable material and 3: unused surface, hide and antler (ibid, 261). This suggests that even when the tools are used for the same amount of time the resulting polishes overlap. The inclusion of antler (a medium to hard material) with unused surfaces and hide (a soft material) is a curious anomaly. Apart from antler, the three groups represent soft, medium and hard materials and the resulting differences in texture probably reflect the sequence of polish development described in Chapter 4 Section 6.9, the three groups corresponding to A, B and C levels of polish development when the duration of tool use is held constant, as in this case.

The model of polish formation presented in Chapter 4, Section 1, implicitly accepts that when tools are used for the same period of time on such different materials as bone and hide, then the resulting polish textures will be distinctive because of the difference in hardness of the worked materials. The conclusion that "variation in flint surface texture caused by flint tools being used on different raw materials ... may be separated by the use of applied texture measures" is based on tools used for the same length of time. As the authors remark "variations in surface texture caused by variable material conditions and time of use are experimentally documented. These parameters are, however, not considered here" (ibid, 258). They are aware of these parameters, however, as they say " It might well be that tools used for longer or shorter periods of time develop surface textures which will give rise to a stronger amalgamation of texture groups" that is, amalgamation of texture groups that include tools used on different materials.

The second part of the texture analysis was to classify a sample of archaeological flint textures using the discriminant functions derived from the experimental design set. The results of this classification were then compared with the results produced by standard optical microwear analysis. A criticism of this procedure is that in order to test the discriminant functions derived from the design set, another sample of experimental tools should have been used, thus comparing images produced from known uses with the design set classification rather than archaeological tools whose precise uses are a matter of interpretation. The normal procedure in such an experiment is to compare the classification produced from the design set with a similar but different test set before applying the classification procedure to archaeological tools. For example the image processing described here derived a model of an unused flint texture from one data set and then analysed another sample of unused flint textures by comparing them with the model (see Figure 20). The same procedure was carried out for polish produced from using flints on antler (see Figure 21). The lack of agreement between the classification from texture analysis and that from visual inspection (ibid, 267; Figure 6), is difficult to interpret because the real use of the tools remains unknown. Also, the superimposition of the archaeological data set onto the canonical variates diagram (ibid, 265; Figure 5a) does not appear to show any clustering of the textures or any association with the groups derived from the experimental design set. The authors of this paper conclude that it would be unsound to use these techniques to identify the materials that the archaeological tools were used on. They further conclude from the analysis of the archaeological tools that "frictional contact with different raw materials for different periods of time have to be accounted for. It can be assumed that this will tend to diminish the differences observed between certain wear types" (ibid, 270), thus supporting the contention expressed here that textures produced by contact with different worked materials for varying periods of time will overlap to the point that they become indistinguishable.

CHAPTER 3:
INVESTIGATING HAFTING TRACES WITH IMAGE PROCESSING

The image processing techniques described above demonstrated that alterations to an unused flint surface could be reliable detected. These techniques have accordingly been used to investigate hafting traces on drill bits (see Unger-Hamilton et al. 1987). In this case the classification had the limited aim of deciding whether any traces were significantly different from unused flint, and of detecting their origins if possible.

The drill bits from the original experiments (made in conjunction with a microwear analysis of drill bits from Abu Salibihk, Iraq) were re-examined microscopically and two cases of some alteration of the flint surface were observed on the hafted areas, but these traces covered such a small area and were so indistinct that no conclusions as to their origin could be reached. Therefore it was decided to carry out a further experimental programme with the specific aim of detecting hafting traces.

The experimental program is illustrated in Figure 26.

Figure 26

DRILL BIT	TYPE OF HAFT	MATERIAL DRILLED	DURATION
1	*loaded* resin	shell	10 minutes
2	wedged	shell	10 "
3	wedged	stone	10 "
4	*loaded* resin	ivory	10 "
5	wedged + chocks	turquoise	5 "
6	sinew bound	malachite	7 "
7	sinew bound	malachite	7 "
8	wedged + chocks	lapis lazuli	10 "
9	wedged + chocks	lapis lazuli	10 "

The *loaded* resin haft consisted of resin mixed with ochre and sandy grit in which the drill bit was held entirely by the mastic (Figure 27a). This was achieved by heating the mastic, inserting the drill bit and allowing the mastic to cool and harden. Wedge haftingwas achieved by making a mortise in the end of the drill shaft in which the bit was held in place by the pressure exerted between the drill shaft and the contact material (Figure 27b). This technique is similar to that described by Keeley as *jam* or wedged hafts, and was chosen because such hafts, "characteristically allow movement of the tool in the haft" (Keeley 1982,799), and therefore are more likely to produce hafting traces. A refinement of this technique was to insert small chocks of wood between the drill bit and the shaft to reduce movement in the haft, and allow the drill bit to be centered along the axis of the shaft more accurately (Figure 27c). The forcing of the slivers of wood into contact with the drill bit would perhaps produce characteristic hafting traces. Hafting with sinew was carried out by cutting a deep slot into the drill haft into which the drill bit was inserted and held in place by wrapping the sinew around the haft and the protruding edges of the drill bit (Figure 27d). The edges were allowed to protrude so that they would be in contact with the sinew, therefore any movement of the bit in the haft would produce friction with the sinew. This was to test whether any traces would be produced that were attributable to the sinew. *Sinew polish* from haftinghas been reported by Buller (1983).

Figure 27

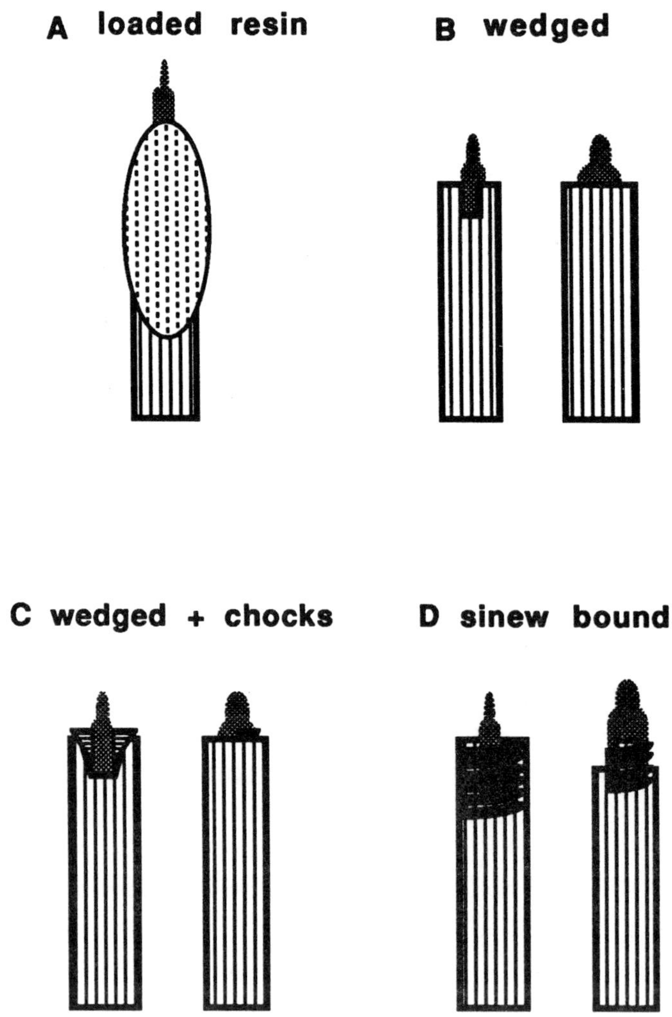

The worked materials were selected as being relevant to the archaeological specimens and the previous experimental program (see Unger-Hamilton et al. 1987).

To attempt to isolate the origin of any traces, micrographs at 200 magnifications were taken of the tools at each technological stage. Photographs were taken of areas of the blanks as soon as they were removed from the core and again after being retouched to form drill bits. The drill bits were then photographed after being hafted and used in the bow drill. Outline drawings of the blanks and the drill bits were made so that the position of each photograph could be recorded. It was attempted to photograph the same area on the blanks and the drill bits, both before and after use, to produce exactly comparable photographs. This was not always possible as some blanks required extensive retouch to reduce them to drill bits and in some cases removing the photographed area of the blank. All areas showing any visible alteration of the flint surface at 200 magnifications were photographed, after the drill bits were retouched, and again after they had been hafted and used. A 100 photographs were taken to obtain as complete a visual record as possible, so that any wear traces observed could be assigned to the technological stage at which they appeared (see Figure 28).

Figure 28

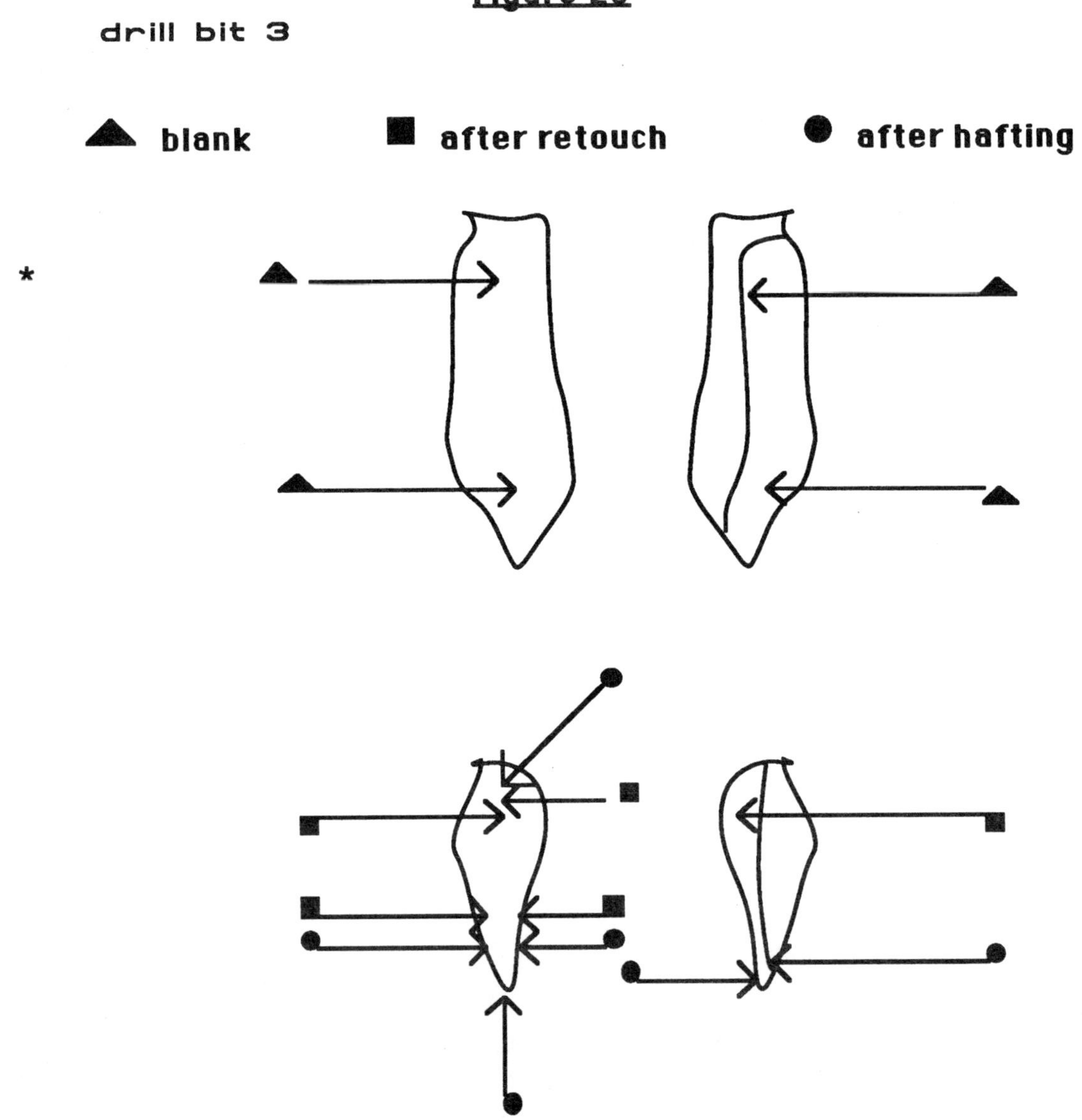

A number of these photographs were selected for digital scanning and subsequent texture analysis, the results of which are illustrated in Figure 29. The values of the statistics are expressed as differences from unused flint so that the nearer a point is to the origin the closer the similarity of that texture sample is to unused flint as described above.

Figure 29

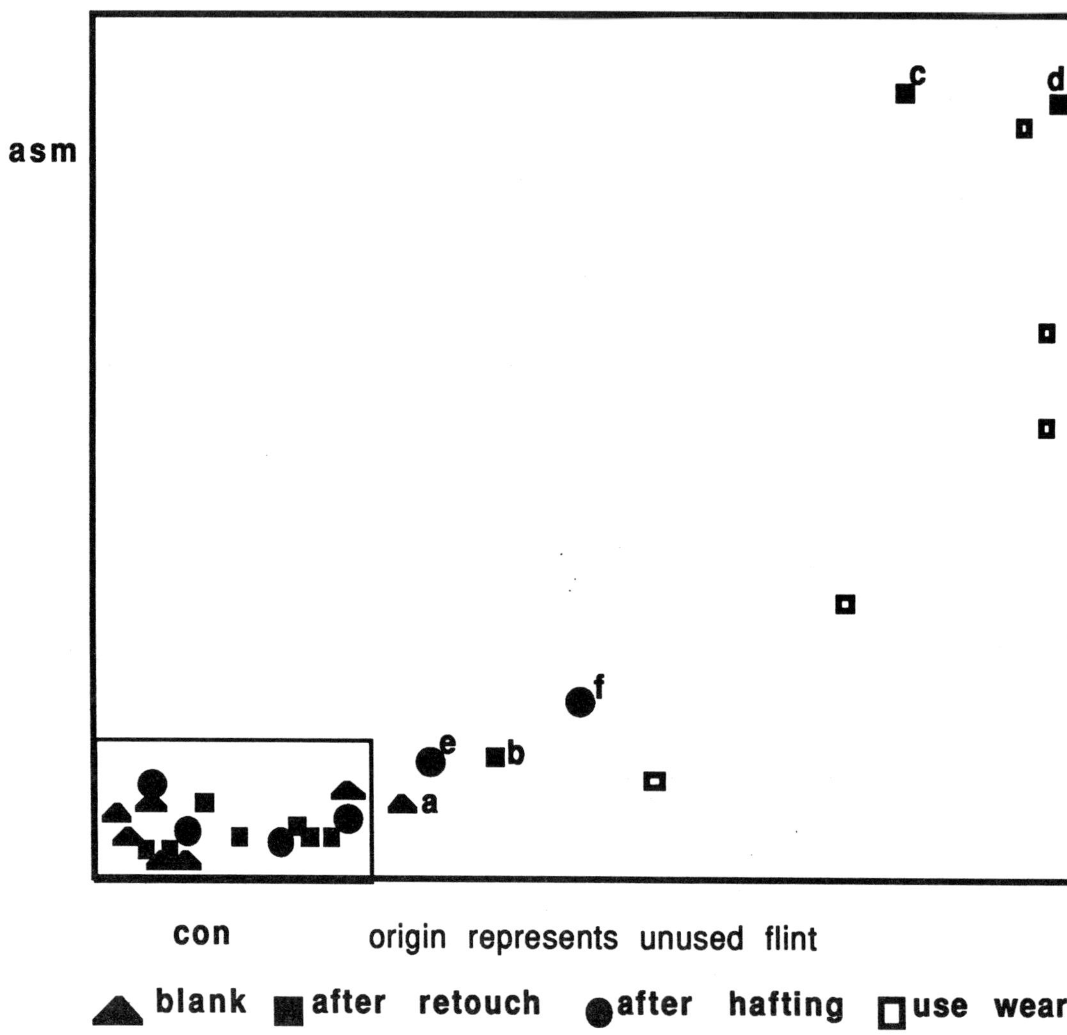

As can be seen from Figure 29 most of the samples cluster around the origin indicating their similarity to unused flint, and thus exhibiting little surface alteration. This group includes textures of blanks, retouched drill bits and used drill bits from photographs of the same area of the same tool. This illustrates that in most cases no surface alteration occurred either from retouch or hafting. Included in the diagram (Figure 29) are five examples of photographs taken of use wear produced at the tip of the drill bits. The area marked around the origin of Figure 29 represents 15% of the total range of variation from unused flint. Allowing for the natural variation of flint textures, points within this area are not considered significantly different from unused flint. This level of significance is calculated from the results presented in Figure 20. Six points representing areas on the hafted end of the drill bits remain outside this area.

Point **a** is from a photograph taken of the bulbar scar of a blank, and the slight surface alteration may be due to the pressure exerted between the core and the blank when it was struck. Points **b, c** and **d** are from photographs taken of retouched drill bits before they were hafted. In all three cases these areas were on dorsal ridges of the drill bits. A possible explanation is that as the tools were pressure retouched on a hide pad, small chips of flint were removed that congregated beneath the tool, so that when pressure was applied during retouch the drill bits were pressed down onto the flint chips causing flint on flint contact. The dorsal ridges would provide a particularly vulnerable area for surface alteration under these circumstances (see Plate 6a) and it is interesting to note that traces on dorsal ridges have been interpreted as evidence for hafting by Keeley (1982 Fig.1 and 5, 805), and Frison (1968,152).

The incidence of such retouch traces prior to hafting on the drill bits could be considered more probable because these tools are retouched around most of the periphery, but the retouching of the non-utilised end of hafted tools of other types is more extensive than with unhafted tools. The shaping of endscrapers to fit the haft, or the backing of bladelets for hafting as knives or sickle blades are examples of this.

Points **e** and **f** are from used drill bits. Point **f** is also from a dorsal ridge (see Plate 6b) of the hafted end of drill bit number 2; **e** is from the central part of the ventral surface of drill bit number 9. Neither of these examples has a polish that is sufficiently distinctive to identify any contact material that may have caused it, and therefore cannot be matched to any particular type of haft. Number 2 was wedge hafted and number 9 was wedge hafted with the addition of chocks, so that if these surface alterations were due to hafting they should exhibit characteristics that are attributable to contact with wood. Not only do they not do so, but the two surface alterations are quite different.

From 9 drill bits hafted by various techniques, only two small and indistinctive areas of surface alterations were observed that occurred after hafting. A far higher incidence of positive hafting traces would be expected to appear on 9 drill bits if hafting traces occur with any frequency, especially considering the intensity of use and the amount of friction between the tool and haft involved in using drill bits in a bow drill. The drilling action involves considerable vertical pressure applied on the capstone as well as the vigorous rotational movement of the drill. The rotational action of the drill takes the form of alternative clockwise and counter-clockwise movement causing lateral vibration as the inertia of the previous movement is overcome.

The fact that surface alteration occurred on the retouched drill bits before hafting indicates that there may be other origins for traces that may have been attributed to hafting. Wear traces from the manufacture of tools has been reported before by Keeley (1980) and Moss (1983). Antler traces from the soft hammer rubbing over the surface of the tool during retouch have been observed (Moss 1983 see Figure 6.8d, 103). But when the manufacturing traces are indistinct and not identifiable to a known process, perhaps there has been a tendency to ascribe these traces to hafting because they appear on the assumed hafted area of the tool. In this study surface alteration occurred on dorsal ridges of tools during retouch, although those areas are well away from the edge where such traces as hammer stone stripe (Keeley 1980) might be expected.

The results of these experiments show that wear traces do occur during retouch and possible minor surface alterations can occur when the blank is removed from the core. Haft wear has been defined as wear traces "which make little sense as traces of utilisation, but does conform to what is known or expected of wear traces from minor movements of a tool against its haft" (Cahen et al. 1979,681). The surface alterations from the manufacture of these tools fall within this definition and could easily be confused with genuine haft wear.

With prehistoric material the tools can only be observed after all the stages of manufacture, hafting and use have taken place, and so it is impossible to be precise in assigning the traces to a particular stage in the process. By careful observation of this process it has been demonstrated that wear traces on the non-operational end of a tool are not necessarily due to hafting and it is suggested that extreme caution should be applied before attributing any traces to hafting.

CHAPTER 4:
A MULTI-DIMENSIONAL APPROACH TO FUNCTIONAL ANALYSIS

SECTION 1:INTRODUCTION

SECTION 2:CLEANING

SECTION 3:MICROSCOPY

3.1 SPECIMEN MOUNTING
3.2 MICROSCOPES
3.3 PHOTOGRAPHY

SECTION 4:OBSERVATION RECORDING

SECTION 5:THE VARIABLES

5.1 RAW MATERIAL
- 5.1.1 GRAIN SIZE
- 5.1.2 TOPOGRAPHY
- 5.1.3 TOPOGRAPHIC FEATURES

5.2 EDGE MORPHOLOGY
- 5.2.1 EDGE ANGLE
- 5.2.2 EDGE LENGTH
- 5.2.3 THICKNESS
- 5.2.4 EDGE PROFILE
- 5.2.5 SHAPE

5.3 EDGE WEAR
- 5.3.1 FRACTURES
- 5.3.2 FRACTURE TYPES
- 5.3.3 ROUNDING
- 5.3.4 GLOSS

5.4 MICRO-EDGE WEAR
5.5 MICRO-TOPOGRAPHY OF POLISHED AREA
5.6 POLISH DISTRIBUTION
5.7 POLISH DISTRIBUTION TYPES
5.8 INVASIVENESS
5.9 LINEAR FEATURES
5.10 STRIATIONS
5.11 POLISH DEVELOPMENT

SECTION 6:FUNCTIONAL SIGNIFICANCE OF VARIABLES

6.1 EDGE MORPHOLOGY
6.2 MACRO-EDGE WEAR
6.3 MICRO-EDGE WEAR
6.4 MICRO-TOPOGRAPHY OF THE POLISHED AREA
6.5 POLISH DISTRIBUTION
6.6 POLISH DISTRIBUTION TYPES
6.7 INVASIVENESS
6.8 LINEAR FEATURES AND STRIATIONS
6.9 POLISH DEVELOPMENT

SECTION 1: INTRODUCTION

The basic assumption behind the technique of high power microwear analysisis that contact between the flint tool and a specific material like wood, will produce a polish which is distinct from that produced by contact with some other material like bone. This assumption has come into question (Newcomer, Grace and Unger-Hamilton 1986). The failure to identify specific wood, bone, antler etc. polishes in blind tests (Unrath et al. 1986), has demonstrated that, rather than being distinctive, these polishes overlap to the extent that they represent a continuum, not discreet recognisable entities. This lack of distinctiveness has been further demonstrated by texture analysis, using image processing techniques, which involves the measurement of the differences between polishes created on tools by contact with various materials by mathematically characterising the textures of the polishes. These techniques and the results of the texture analysis were presented in the previous chapter.

In response to the fundamental doubts that have been raised about the accuracy and usefulness of microwear analysis as it is currently practised, a new approach which attempts to standardise the methodology of microwear analysis and to test the limits of its interpretations has been developed. The method involves the systematic recording of the functionally diagnostic attributes of a tool. These attributes are described using a standard vocabulary and the descriptions can be replicated enabling different analysts to describe the same tools in similar ways. Correlations between the variables then allow the analyst to eliminate some of the possible functions of a tool until the most probable function is isolated. In some cases the elimination of possible functions leaves only one that is consistent with all the wear traces on the tool. This means that functional reconstructions which include the specific material the tool was used to work can be postulated with some confidence.

Other analysts describe polish characteristics in terms of smoothness, brightness and such features as *comet shaped* pits, claiming that there is general agreement about the characteristics of particular polishes. However a comparison of different workers' characterisation of the polish left by working bone, for example, show considerable disagreement; one describes bone polish as bright, rough with micropits less than 1 µm. in diameter (Keeley 1980,43), another as rough with depressions between 10-20 µm. in diameter (Anderson-Gerfaud 1981 v.1, 58). Further descriptions of bone polish are matt, bright, rough or smooth with depressions (Moss 1983, 92), and finally as bright, rough or smooth with micropits, depressions and *comet shaped* pits (Vaughan 1981, 140). Thus bone polish can be rough or smooth, matt or bright, with micropits or depressions or both, and there seems to be no real agreement about the precise appearance of a distinctive bone polish. Gendel and Pirnay (1982) state that their *smooth* bone polish clearly resembled

Keeley's *smooth* antler polish, and so made no attempt to distinguish between bone and antler as worked materials.

Microwear analysts do use information other than polish characteristics in their functional reconstructions of tool use, but this usually takes the form of supportive evidence for determining the motion of the tool. Gendel and Pirnay (1982) in their publication of the results of a blind test, state that they rely principally on the distinctive nature of polishes but also that "Inferring the method of use involves the consideration of several factors other than the distinctive quality of the microwear polishes. These include the morphology of the working edges, utilisation damage, the orientation of striations and the location and extent of microwear polishes" (ibid, 252). They do not however, present this information is a detailed or systematic way. In addition they note specific features as diagnostic of worked materials. For example, the two dry hide working tools in their test were identified by the presence of rounded edges and short striations. In experiments conducted at the Institute on dry hide these features are not always present and so cannot be assumed to have a direct correlation with a specific material. The approach to use-wear analysis employed here does not rely on any single variable being diagnostic, but on the agreement of all the variables which lead to a logically consistent functional reconstruction. proceeds on the assumption that distinctive polishes are produced by working different materials, hence bone polish, hide polish etc. The approach presented here is based on a model of polish development as a continuum, and no attempt is made to assign a polish as material specific.

The theory behind polish development began with a model of distinctive, material specific polishes (Figure 30). This has been demonstrated to be too simple by the repeated failure of analysts to recognise these *distinctive* polishes consistently and so the model has been modified. It has been accepted that all materials produce the same polish during the early stages of tool use, and then as work progresses the polishes develop separately, though overlapping does occur as in Figure 31 (after Vaughan 1985, 46).

Figure 30

A: as separate entities

Figure 31

overlapping

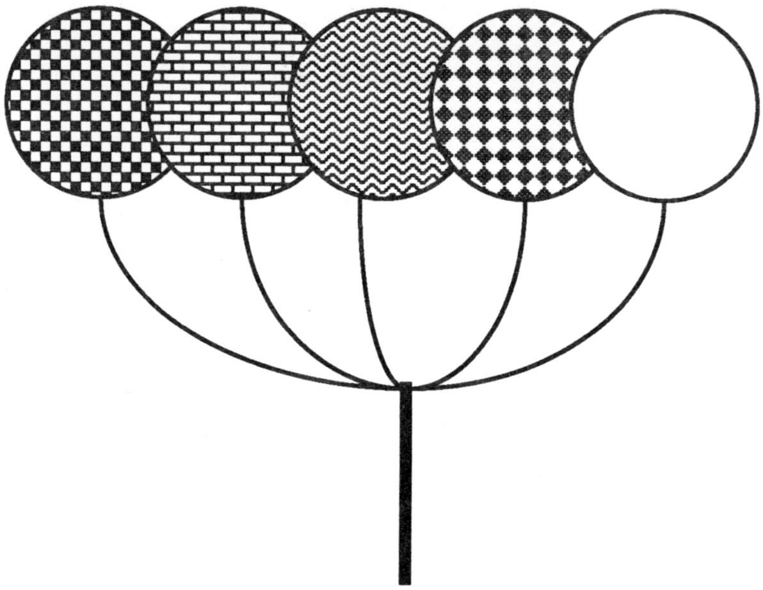

The failure in blind tests to discriminate between these polishes (Newcomer et al. 1986, Unrath et al. 1986) and the results of the texture analysis of microwear polishes (Grace et al. 1985) indicates that the real situation is that the overlapping of polish types is far greater, as illustrated in Figure 32.

Figure 32

realistic overlapping

Consequently a model of polish development as a continuum has been adopted. The important variables of this model are the hardness of the worked material and the duration of use, therefore the model illustrated in Figure 33, is proposed.

Figure 33

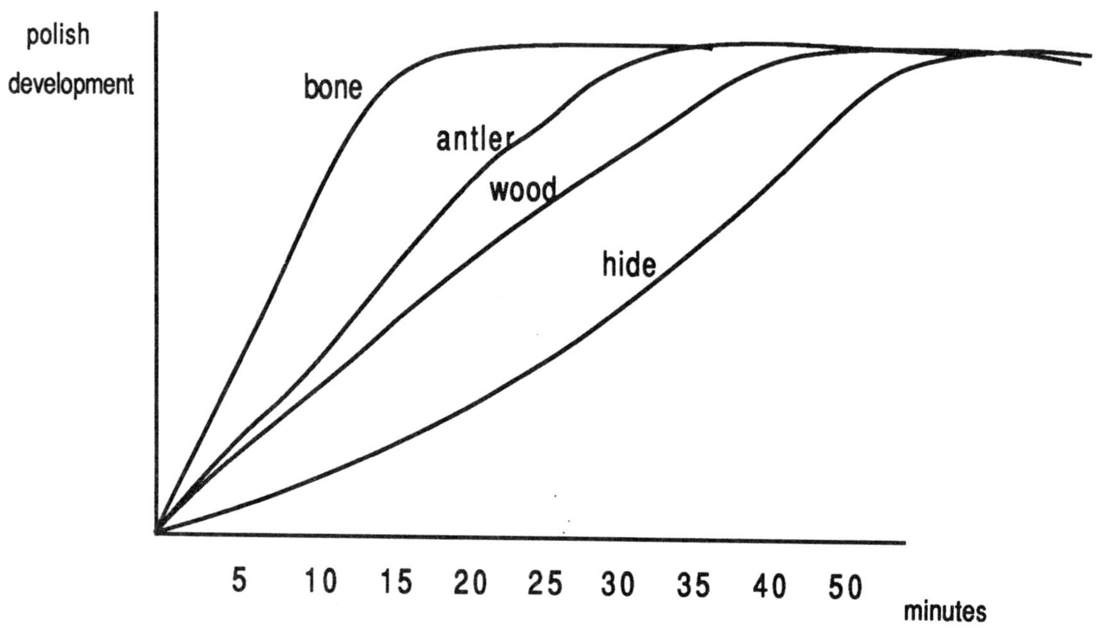

In this model polish develops as a continuum until the used area is completely polished with no unused flint texture remaining. The *shape* of the curve is determined by the hardness of the worked material so that flints that have been used on bone will develop polish more rapidly that polish produced by working a softer material such as hide. This means that a particular level of polish development observed by an analyst (such as **X** in Figure 34) may be produced by, for example, use on bone for 6 minutes, antler for 11 minutes, wood for 14 minutes or hide for 27 minutes.

Figure 34

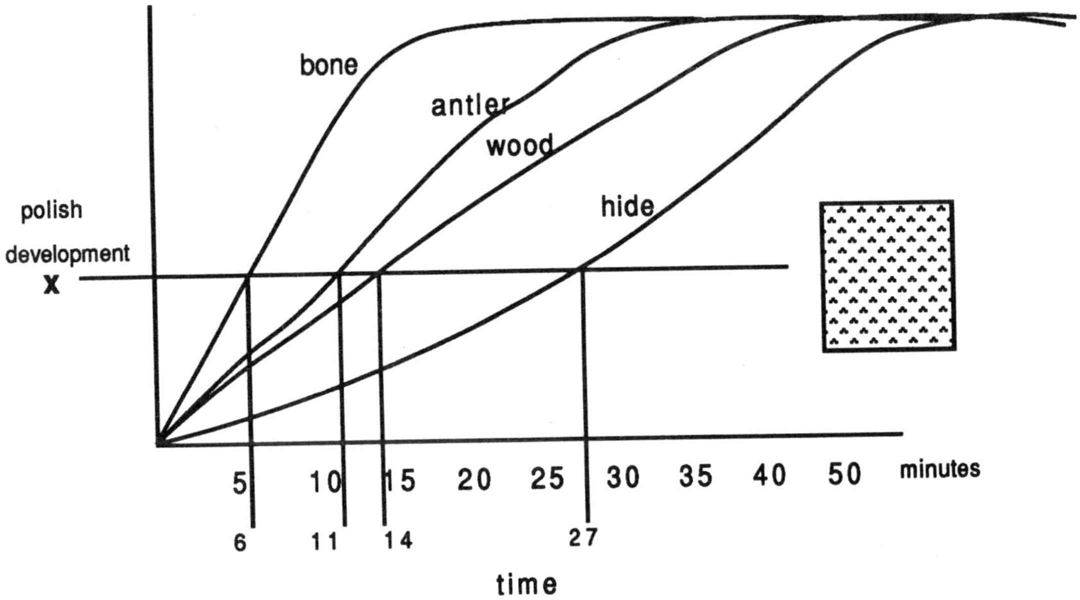

The fact that the duration of tool use is unknown with archaeological material, means that polish X could be the result of working any of these different materials. Therefore the model used in conjunction with this approach is to treat polish as a single continuum as in Figure 35.

Figure 35

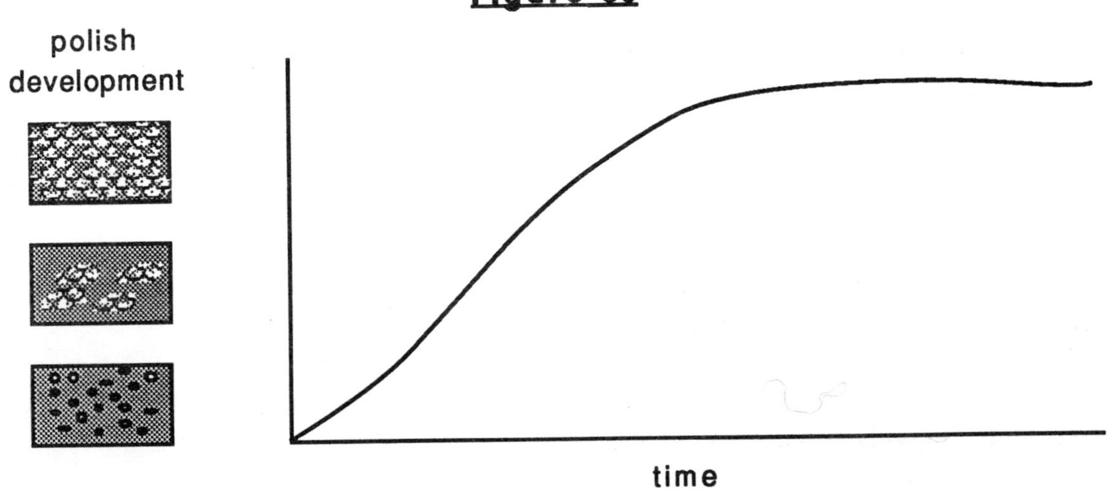

The reasons for adopting this model are that until more research has demonstrated the precise mechanism of polish formation and development, it is preferable to use the simplest explanation that concurs with the observed data (employing Ockham's Razor). If further research allows for a more precise model that is helpful in interpreting the function of stone tools then the model will be modified accordingly.

The following explanation of the multi-dimensional approach to use-wear analysis is divided into sections that deal with the different aspects of the method.

SECTION 2 deals with the procedures for cleaning the flints used in conjunction with this method.

SECTION 3 deals with the practical aspects of using microscopes for use-wear analysis and the photographic recording of polish images.

SECTION 4 explains how the observations are recorded.

SECTION 5 defines the variables and their values and describes how the values are measured.

SECTION 6 discusses the functional significance of the variables. The definition and measurement of the variables are separated from the discussion of their functional significance so that SECTION 5 may be used as a manual for observation recording. It is important that observations are separated from interpretation. If these two aspects are not carefully delineated they can often be confused, so that the reader is not sure if he is dealing with an observation or an interpretation of an observation.

SECTION 7 describes the process of interpretation and the way in which attributes are combined to eliminate motions of use and worked materials until the most probable function can be interpreted. Also, Section 7 includes the definitions of the different motions of use that are employed in this method.

SECTION 8 gives examples of tools whose function have been interpreted by this method in blind tests.

SECTION 9 describes the role of experimental replication of the function of tools that is an integral part of this method.

SECTION 10 describes the blind test used to assess the usefulness of this method and contrasts it with previous blind tests.

GLOSSARY

SECTION 2:CLEANING

The cleaning procedure used with this method is to wash the tools with detergent and water to remove dirt or soil deposits. To remove deposits still adhering to the tool an ultra-sonic tank is used for 10 minutes with the tool immersed in distilled water. The ultra-sonic tank will remove such things as soil particles embedded in hollows in the tool. Caustic chemicals advocated by Laurence Keeley (Keeley 1980, 3) and some other analysts, are not used for cleaning for the following reasons.

1) Microwear analysts have compared prehistoric tools that have been chemically cleaned with those that have not, and found that the polishes are not significantly changed (Moss 1983, 105).

2) Experiments using the least caustic of these chemicals (**NaOH**) for only 10 minutes in a dilute solution, have shown that even such relatively mild chemical treatment may visibly alter the surface of an unused flint.

3) The question of how much polishes are made up of deposits of the worked material has not been definitively answered (see Vaughan 1985), so that deposits, for example of bone, may be one of the constituents of polish, and chemical cleaning will remove this. Keeley even suggests that the change in the appearance of bone polish after chemical cleaning is due to the removal of bone apatite (Keeley 1980, 43). If such deposits are preserved on the tools it would be preferable to develop techniques of residue analysis to identify positively the worked material. The recent investigations into the use of blood residue analysis (Loy 1983) demonstrate that if such deposits do survive, chemical cleaning could be destroying potentially valuable information. Materials such as antler and bone, if they survive at a site and they are a constituent of polish, will also survive on tools and be susceptible to residue analysis.

If chemical cleaning is not used there is the problem of organic residues on experimental tools that remain even after washing and the use of the ultra-sonic tank, residues produced by the working of green wood are particularly troublesome and remain in some cases even after the tool has been cleaned in white spirit, which removes mastic used for the hafting of tools. This problem is particularly relevant to tools used in blind tests, as the presence of a residue from wood working could enable the analysts to make a correct identification of the worked material from evidence that would not survive archaeologically. To overcome this problem experiments with biological cleaning agents have been carried out and it was found that ordinary household washing powders that contain enzymes remove such organic deposits very efficiently. Simple immersion for 10 minutes in a solution of a biologically active washing powder removes not only residual deposits, but also cleans off such things as grease and any plasticine that may have adhered to the tool when it was mounted for observation through the microscope.

Cleaning of the tool's surface during microscopic observation is carried out by wiping the surface with a cotton bud dipped in alcohol. Periodic cleaning with alcohol is necessary for removing deposits such as finger grease that are produced from handling the tool. The use of cotton buds can sometimes produce *false striations*, when grease or remnants of plasticine are present on the tool. The wiping action can spread such residues in one direction giving the appearance of linear features, and so it is necessary to be aware of this. A rotational action of wiping avoids this confusion as if such false features are produced they become obvious from such a rotational action. If such residues are shown to be present on the tool, it should be removed from the microscope and re-cleaned following the procedure described above.

SECTION 3:MICROSCOPY

3.1 SPECIMEN MOUNTING

Plasticine is used to mount the tool on a specimen slide. The main disadvantage of plasticine as a medium for mounting the tools is that it will adhere to the tools, particularly in hollows, therefore requiring cleaning as described above.

3.2 MICROSCOPES

The microscopes used at the Institute are the **Olympus Vanox** and the **Olympus BMHJ**, the latter being a portable microscope more suitable for field work than the larger Vanox model. Both models are metallurgical microscopes using bright field illumination and have the same optical system. The stages move in two directions using separate controls that can be manipulated easily to facilitate the scanning of the tool. The magnifications available are 50x, 100x, 200x, and 400x. Magnification of 100x is mainly used for scanning the tools and when any features are located they are examined at 200x. The magnifications of 50x and 400x are only used occasionally. 50x magnification is useful when the tool's surface is so irregular that at 100 or 200 magnifications very little can be observed because of the lack of depth of focus at those magnifications. All microscopic recording of variables are made at 200 magnifications.

3.3 PHOTOGRAPHY

The Olympus Vanox has a photographic system attached that allows the exposure to be set automatically by measuring the light level through the optics of the microscope. The Vanox has in built filters; only the yellow filter is used in order to enhance contrast. The BHMJ model has a simple adapter tube that allows for an ordinary SLR camera with auto-exposure to be attached.

Ilford FP4 film rated at 125 ASA is used but for better results **Pan F** film rated at 50 ASA is preferred. To reduce the exposure time Pan F is up rated to 125 ASA and developed with Acutol which increases the exposure by one stop. By developing for double the recommended time the process of under exposing and over developing produces a clearer image with more contrast. To enhance this contrast grade 4 photographic paper is used to produce prints. Even with these techniques it is often difficult to produce high quality photographs because of the *greyness* of the images. Also the small depth of focus at 200 magnifications means that is often difficult to achieve focus over the whole of the negative. Photographs taken at 200 magnifications represents 625x425 µm. of the flint surface. When the size of the image is too small to illustrate a particular feature 100 magnifications may be used which represents 1250x850 µm. of the flint surface. Occasionally, to illustrate larger features such as fractures, 50 magnifications may be used which represent 2500x1700 µm. area of the flint surface.

SECTION 4: OBSERVATION RECORDING

An outline drawing is made of the tool on which is recorded the position of any features and also the location of any photographs that might be taken. It is not necessary to make a detailed drawing of the tool, a schematic representation of the retouch being sufficient rather than drawing individual flake scars (Figure 36). Other workers use polar coordinates to record the location of used areas, but when illustrating tools the worked edges are indicated on a drawing (see Odell and Odell-Vereeken 1981). The remaining recording is carried out by filling in the values for the variables on the observation sheet (Figure 37). The main purpose in using an observation sheet is to enable the analyst to record the variables in a systematic way and to keep the observations consistent between one tool and the next. Also, the presentation of information in this form provides primary data which can be referred to by other analysts when assessing any functional reconstructions made from the data. Most microwear reports provide polish descriptions using subjective terms which may be interpreted in different ways by different analysts reading the reports. In addition, the cost of publishing often restricts the number of photographs which can be published in a report, thus further limiting the primary data available to the reader. This means that the findings of the microwear analyst are accepted or rejected without adequate information.

The reference number of the tool and its type is recorded. If there is more than one working edge on the tool each edge is given a unit number and is treated separately and a new observation sheet is used. For the definition of a *working edge* see Section 5.2.

With a retouched tool the retouch on the surface (usually the dorsal surface) may make observation of edge wear and polish features impossible, in which case the retouched surface will be recorded as retouched and variables for that surface left unrecorded.

Figure 36

OUTLINE DRAWING

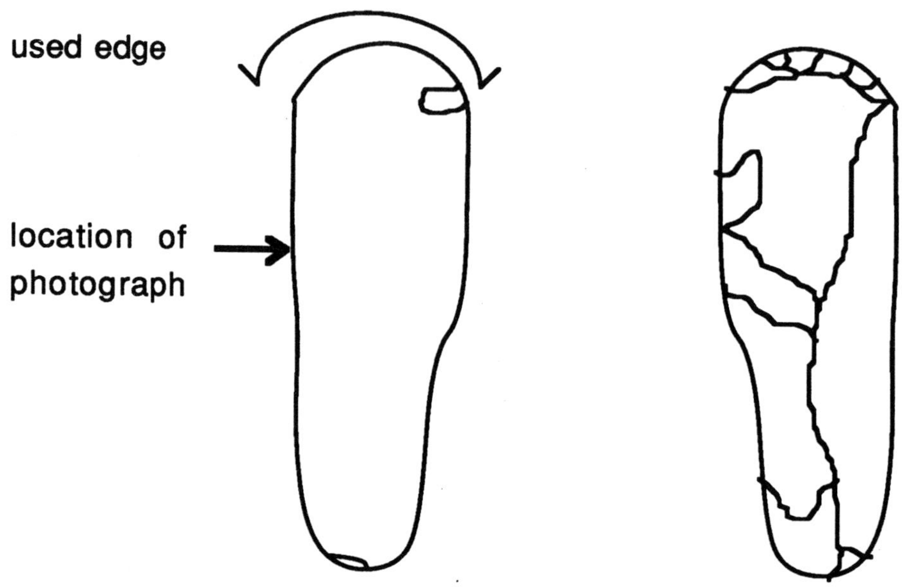

Figure 37

OBSERVATIONS

TOOL NO. **TYPE**

GRAIN SIZE	1 fine	2 medium	3 coarse	
TOPOGRAPHY	1 flat	2 undulating	3 ridged	
TOPOGRAPHIC FEATURES	1 percussion ripples	2 edge feathering	3 both	4 absent

EDGE MORPHOLOGY

EDGE ANGLE

LENGTH

THICKNESS

PROFILE

SHAPE

MACRO EDGE WEAR		**DORSAL**	**VENTRAL**
FRACTURES	1 absent 2 <5 per 10mm. 3 >5 per 10mm.		
FRACTURE TYPE	1 flakes 2 snaps 3 steps 4 flakes+snaps 5 flakes+steps 6 snaps+steps 7 flakes+snaps+steps 8 absent		
ROUNDING	1 light 2 heavy 3 absent		
GLOSS	1 present 2 absent		

MICRO EDGE WEAR

FRACTURES

1 absent
2 <5 per 5 mm.
3 >5 per 5 mm.

FRACTURE TYPE
1 flakes
2 snaps
3 steps
4 flakes+snaps
5 flakes+steps
6 snaps+steps
7 flakes+snaps+steps
8 absent

ROUNDING
1 light
2 heavy
3 absent

MICRO-TOPOGRAPHY OF POLISHED AREA 1 flat 2 undulating 3 ridged

MICRO-POLISH DISTRIBUTION 1 continuous 2 intermittent 3 absent

DISTRIBUTION TYPE
1 away from the edge
2 gapped
3 edge only/even
4 edge only/asymmetric
5 differential
6 absent

INVASIVENESS
1 edge only
2 <0.5D
3 >0.5D
4 absent

LINEAR FEATURES
1 parallel
2 perpendicular
3 angled
4 parallel+perpendicular
5 parallel+angled
6 perpendicular+angled
7 absent

STRIATIONS
1 parallel
2 perpendicular
3 angled
4 parallel+perpendicular
5 parallel+angled
6 perpendicular+angled
7 absent

POLISH DEVELOPMENT
1 A (individual elements)
2 A+
3 B (linked)
4 B+
5 C (all over)
6 D (linear)
7 absent

SECTION 5:THE VARIABLES

5.1 RAW MATERIAL

- 5.1.1 GRAIN SIZE
- 5.1.2 TOPOGRAPHY
- 5.1.3 TOPOGRAPHIC FEATURES

5.2 EDGE MORPHOLOGY

- 5.2.1 EDGE ANGLE
- 5.2.2 EDGE LENGTH
- 5.2.3 THICKNESS
- 5.2.4 EDGE PROFILE
- 5.2.5 SHAPE

5.3 EDGE WEAR

- 5.3.1 FRACTURES
- 5.3.2 FRACTURE TYPES
- 5.3.3 ROUNDING
- 5.3.4 GLOSS

5.4 MICRO-EDGE WEAR

5.5 MICRO-TOPOGRAPHY OF POLISHED AREA

5.6 POLISH DISTRIBUTION

5.7 POLISH DISTRIBUTION TYPES

5.8 INVASIVENESS

5.9 LINEAR FEATURES

5.10 STRIATIONS

5.11 POLISH DEVELOPMENT

SECTION 5:THE VARIABLES

The variables can be divided into five groups; raw material characteristics, edge morphology, macroscopic edge wear, microscopic edge wear and polish distribution and characteristics. These variables and the recording sheet developed for use with this method are illustrated in Figure 37.

5.1 RAW MATERIAL

The raw material is described in respect to its grain size, surface topography and topographic features. These are characteristics which have been shown to affect the appearance of wear traces on stone tools under the microscope, but are not diagnostic of function.

5.1.1 GRAIN SIZE

Grain size is recorded according to the visual appearance of the tool, and often relates to the colour of the stone. For example, the black areas of Brandon flint are usually **fine grained**, the grey areas tend to be **medium grained** and the white area **coarse grained**. This estimation of grain size is made on the basis of comparing flints from different sources and also the grey tone of the flint is often a function of the way in which light is reflected by the differing grain sizes. It is important to record grain size because it can significantly effect the visual appearance of the flint under the microscope. Sometimes the effect is such that coarse grained flint makes the observation of polishes almost impossible, because the level of reflectiveness produces a blurred appearance.

5.1.2 TOPOGRAPHY

An **undulating** or **ridged** topography of the flint surface away from the polished area can affect the distribution of the polish. Therefore the topography is recorded as it may correlate with polish distribution variables, and so indicate that the distribution is not functionally determined, but a product of the original topography of the unused surface. (see SECTION 6.5)

5.1.3. TOPOGRAPHIC FEATURES

These features often explain the origin of the topography. **Percussion ripples** are produced during the conchoidal fracturing of the blank and radiate concentrically from the point of percussion. Such percussion ripples often produce an undulating topography (see Figure 38).

Figure 38

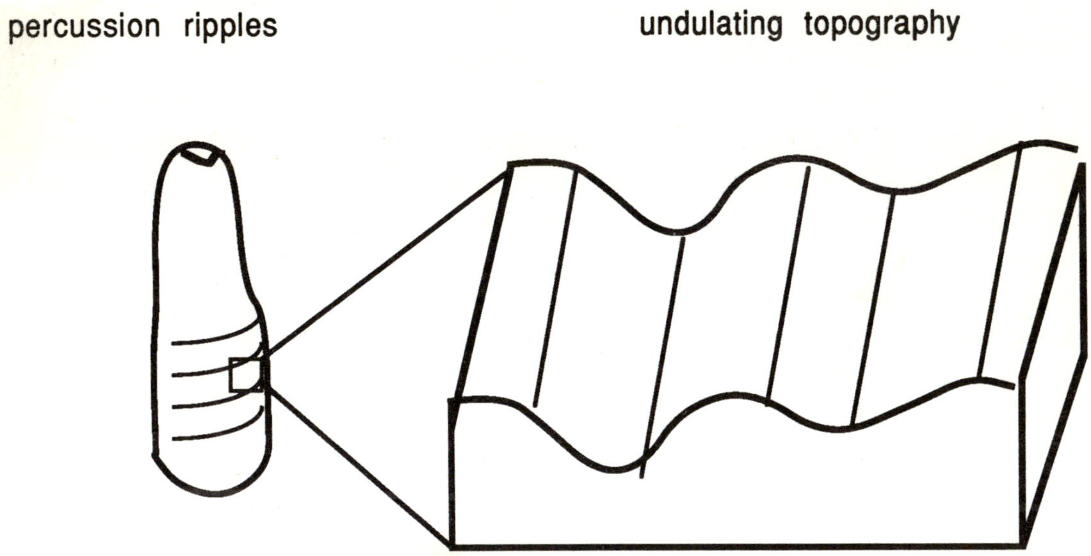

Edge feathering often occurs on thin edges and is a product of the process of detachment of the blank from the core resulting in overlapping fracture planes that have a *feathered* appearance. This often results in a ridged topography on the edges of the tools (see Figure 39).

Figure 39

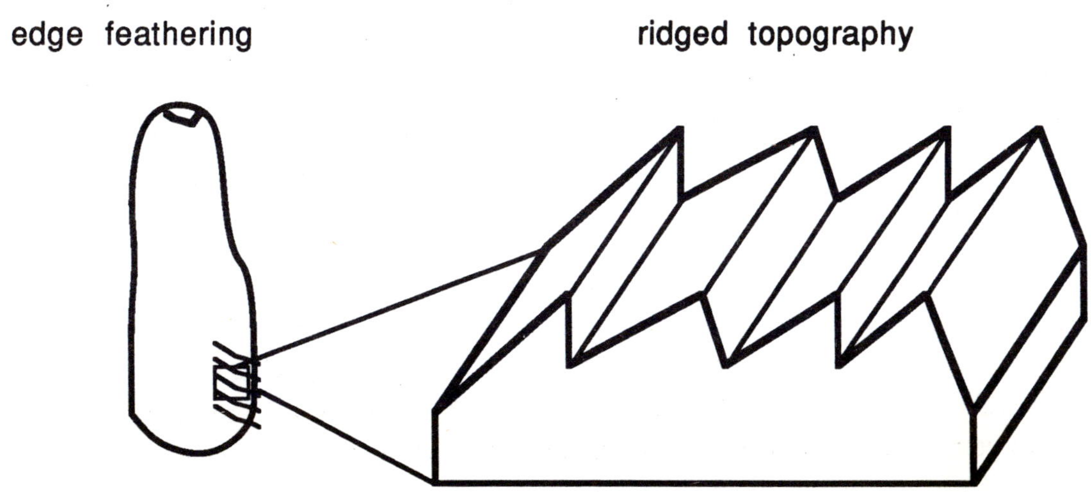

5.2 EDGE MORPHOLOGY

Aspects of the edge morphology are measured metrically and refer to the working edge of the tool. The variables are **edge angle, edge length, edge profile, thickness** and the ***shape*** of the tool. These variables were chosen from a larger group of possible variables, because in a previous study these particular variables were the ones which seemed to provide reliable functional information (Grace 1981). Potentially used edges are identified according to the following definition. Used edges are created by retouch or utilisation or may be natural edges opposed by backing retouch, as with a backed knife.

The ends of the edges are determined by the limits of the retouch or utilisation, or where a continuously retouched edge turns through a sufficiently large angle to indicate the cessation of the worked edge. The criterion of a change of direction of the edge line of more than 40 degrees within a distance of 1 millimeter is used, as suggested by White (1969, 24).

Retouched or utilised edges on the same piece are considered in two ways: one, as lateral (to the used edge) retouch for shaping the *handle*, or manufacturing traces (eg. a truncation or a stop notch on a burin) and two, as distinct and separate working edges. This procedure is necessarily somewhat subjective, but the nature and placement of backing or piece-modifying retouch is usually sufficiently distinctive from retouch designed to produce a working edge. Unretouched pieces that have very little edge damage so that the evidence of utilisation is marginal, can be examined by looking at the relationships between the morphological attributes of such edges. The morphological variables can be used to estimate the functional capability of edges, and can be used not only for functional interpretations (see section 6), but also to operate as feedback to the initial problem of determining whether a piece could have been used or not. For example, the bulbar end of a flake could be described as a short, obtuse angled, thick, convex edge of a short flake, and would not normally be an edge suitable for use.

Pointed tools require some explanation as they represent an exception to the above criteria. A subjective assessment has to be made as to whether the converging retouched edges represent two distinct working edges, or the convergence was deliberately produced for functional reasons; i.e., the convergence was designed to form a point. This problem corresponds with the typing of a tool as a Mousterian point or as a convergent scraper by the thickness of the cross section of the tip (Bordes 1961). Bifacially retouched points, where the retouch is invasive are more obviously *designed* as points. The overall morphology of a projectile point, for example, is indicative of the tool's potential function. The morphological attributes of points are measured slightly differently to other tools. The working edge is taken as both of the edges that converge to a point, so that the middle of the working edge is the apex of the point.

Worked edges thus isolated are marked on the drawing and taken as the base from which to record the morphological attributes. This procedure allows for the selection of potentially worked edges prior to the microscopic examination. Microscopic examination is then used to confirm which edges were actually used, by the presence of micro-edge wear and polish. Though all edges and surfaces of the tools are examined microscopically, the above procedure has proved reliable in isolating worked edges.

5.2.1 EDGE ANGLE

Edge angle is defined as the angle between the ventral surface and the retouched or utilised edge (assuming direct retouch), and corresponds to Wilmsen's edge angle (Wilmsen 1968, 985), and Movius and Brooks angle BAC (Movius and Brooks 1971, 269). Edge angle is not the same as the spline plane angle as described by Tringham et al. (1974). In the case of inverse or bifacial retouch, the edge angle is that of the working edge created by such retouch. Though there can be a variation of the angle along the length of an edge, this is generally slight and measuring the angle at the mid-point of the working edge has been found to be consistently representative (Figure 40).

Figure 40

EDGE ANGLE MEASUREMENT

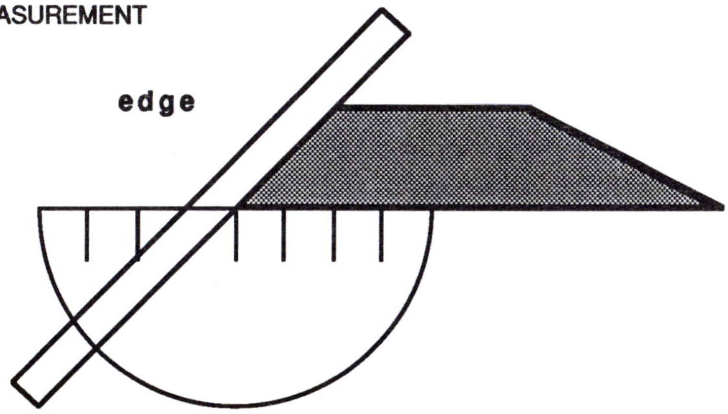

As described above the middle of the working edge of points is defined as the apex of the point, so that the edge angle of points is measured as the angle created by the convergence of the two edges. The edge angle measurements are taken with a goniometer (Figure 41). For a comparative study of edge angle measurement techniques see Dibble and Bernard 1980.

Figure 41

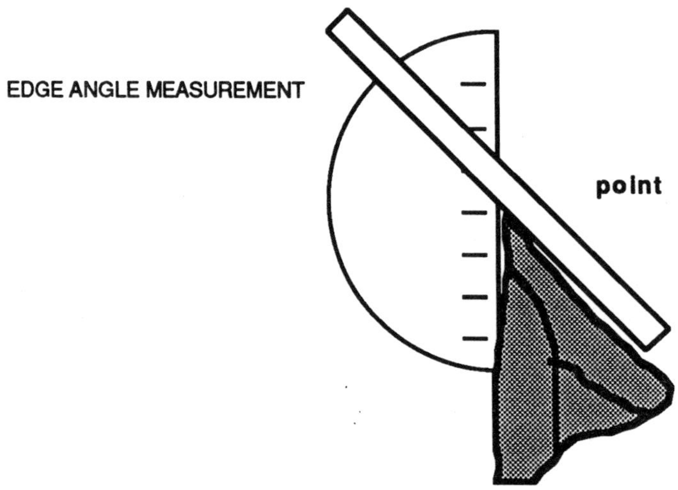

5.2.2 EDGE LENGTH

Edge length is defined as the maximum length of the working edge whether created by retouch, utilisation or a natural edge indicated by backing retouch of the opposing or lateral edge, as explained above. The curvature of the edge is allowed for by fitting a line (using a piece of non stretchable plastic) or a designers curve to the edge. This can then be straightened and measured against a ruler (Figure 42).

Figure 42

EDGE LENGTH MEASUREMENT

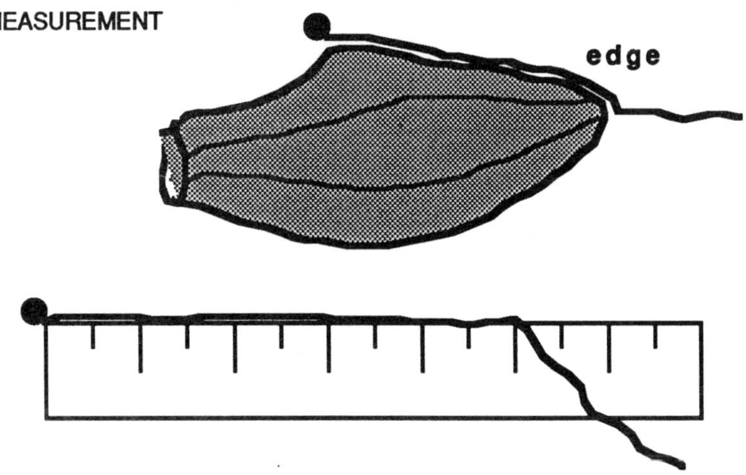

The edge length of points is taken from the *shoulder* round the point to the opposite shoulder (Figure 43).

Figure 43

EDGE LENGTH MEASUREMENT

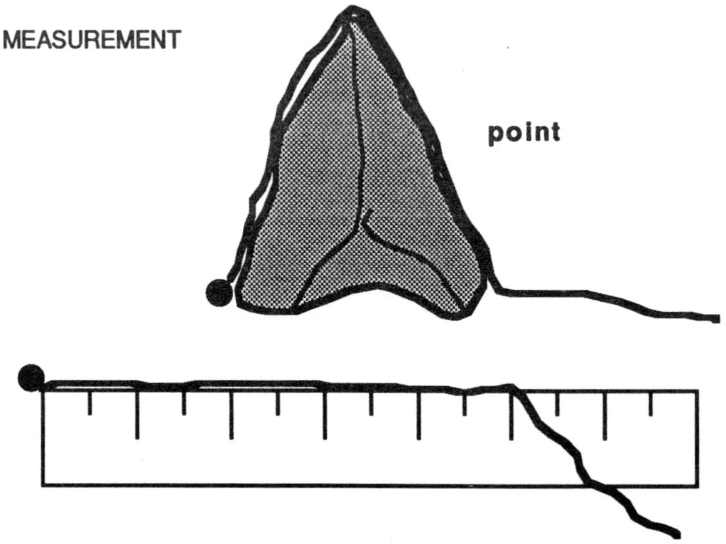

5.2.3. THICKNESS

Thickness is defined as the maximum thickness of the support piece taken perpendicular to the mid-point of the working edge. Measurements are taken with calipers (Figure 44).

Figure 44

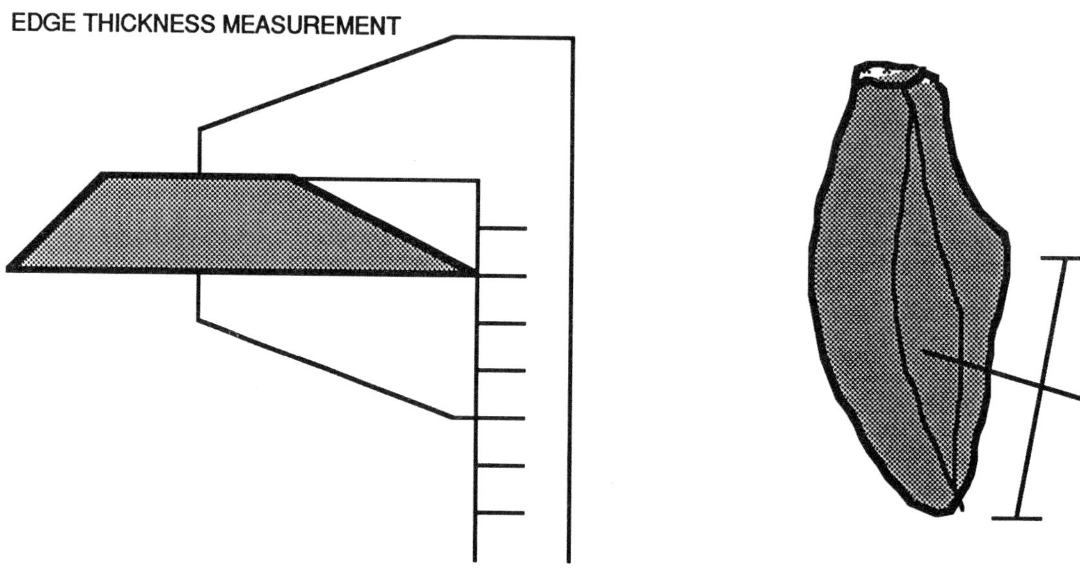

The thickness of points is taken as the maximum thickness of the support piece with the calipers placed at the apex of the point i.e. the mid-point of the working edge as defined here for points (Figure 45).

Figure 45

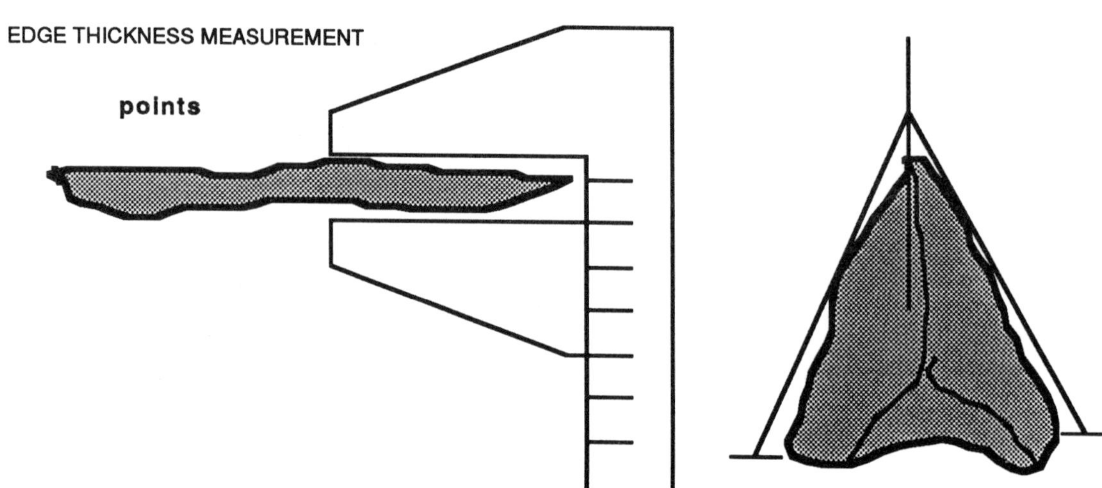

5.2.4. EDGE PROFILE

Edge profile is defined as the shape of the working edge in plan, which may be convex, concave or straight as indicated by the ratio of the perpendicular measurement divided by the chord. The chord is the linear distance between the extreme ends of the working edge. It is measured by aligning the ends of the working edge along the **y** axis of millimeter graph paper and reading off the measurement from the graph (Figure 46).

The perpendicular is measured by taking the maximum distance between the working edge and the **y** axis. The measurement is read off from the **x** axis Concave profiles are given a negative score and by dividing the perpendicular by the chord, straight edges have a value of 0. The profileof points is calculated with the base of the point placed on the **y** axis (Figure 46). For examples of profiles see Figure 47.

Figure 46

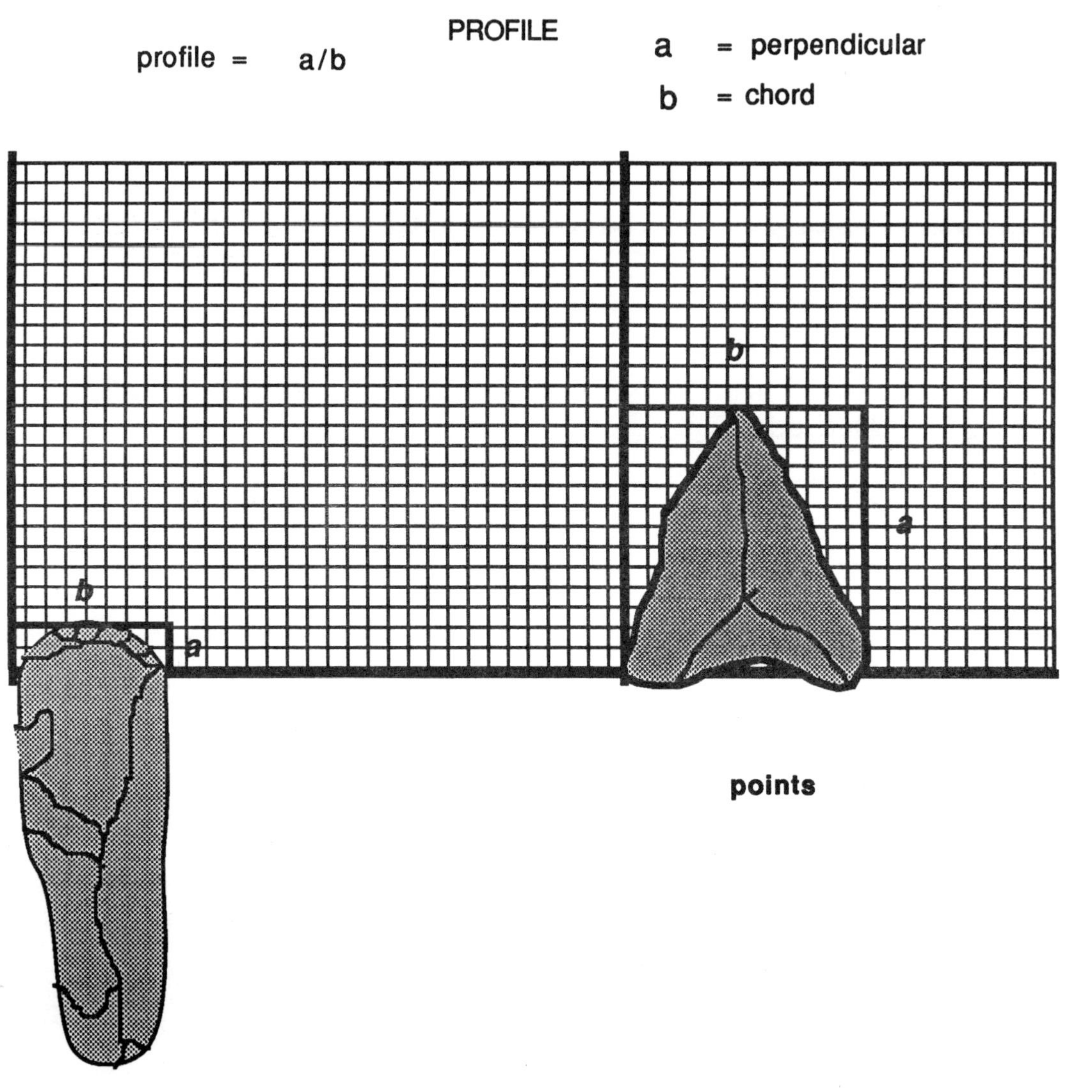

Figure 47

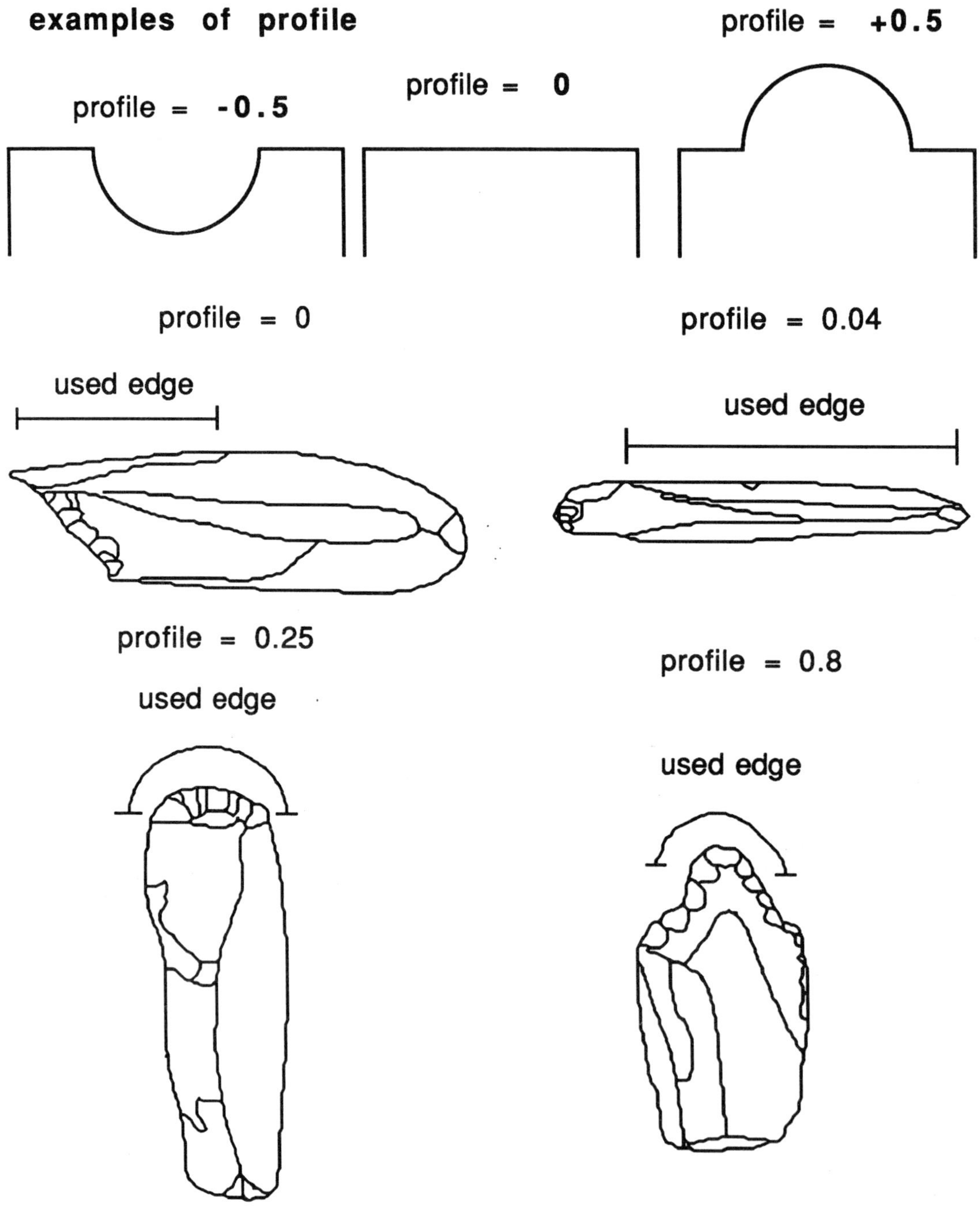

examples of profile
profile = -0.5
profile = 0
profile = +0.5
profile = 0
used edge
profile = 0.04
used edge
profile = 0.25
used edge
profile = 0.8
used edge

5.2.5. SHAPE

The *shape* ratio is calculated by dividing the length by the height. Length is defined as the maximum lateral dimension of the piece with the working edge as the base. It is measured by orientating the working edge along the **y** axis of the graph paper and reading off the value (Figure 48). Height is defined as the maximum vertical dimension of the support piece with the working edge as the base. Measurement is by reading off the value from the **x** axis when the piece is orientated as for length The calculation of the shape ratio means that a long thin support piece, such as a blade supporting an end scraper, will have a relatively low value, compared with a flake supporting a side scraper. A blade with a lateral retouched edge will also have a high value, so that this measurement will not, and is not intended to, distinguish between blades and flakes, as it depends on the position of the working edge in relation to the support piece. It measures the shape of the part of prehension while in its working attitude. The *shape* of points is taken with the apex of the point placed on the **y** axis (Figure 48). For examples of *shapes* see Figure 49.

Figure 48

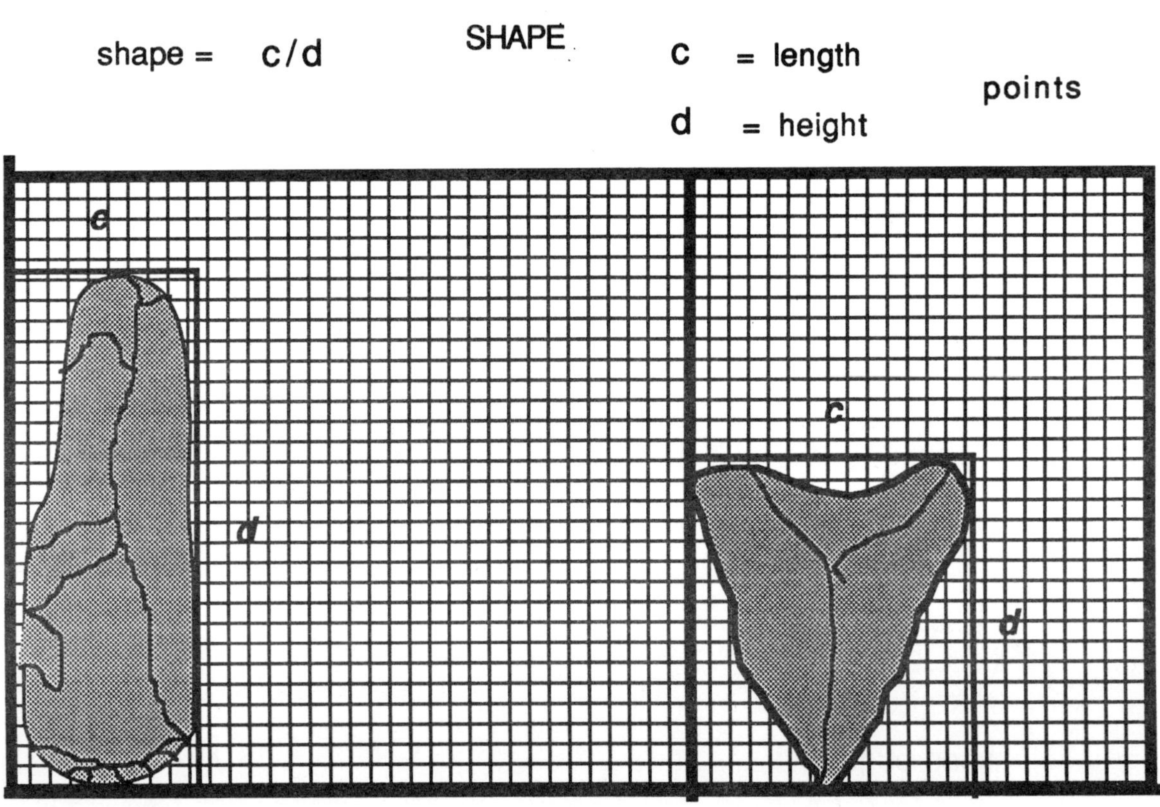

Figure 49

examples of shape

truncated blade shape = 2.24

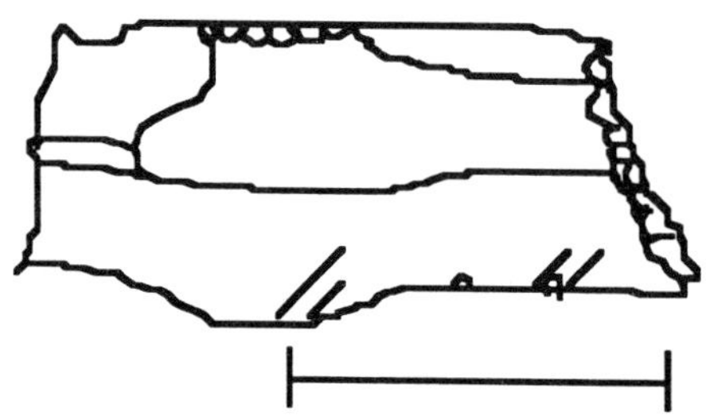

truncation burin shape = 3.9

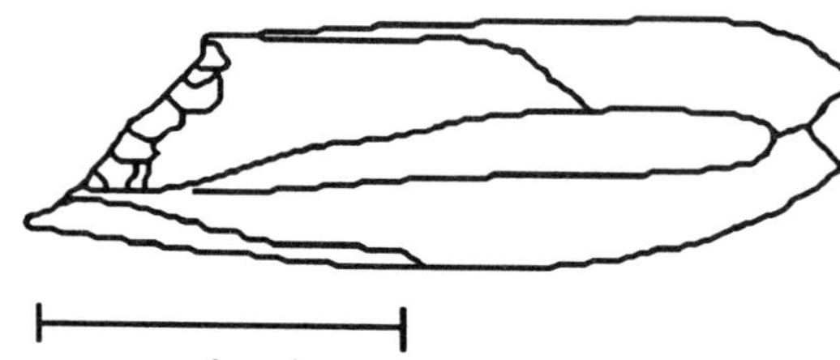

blade shape = 4.57

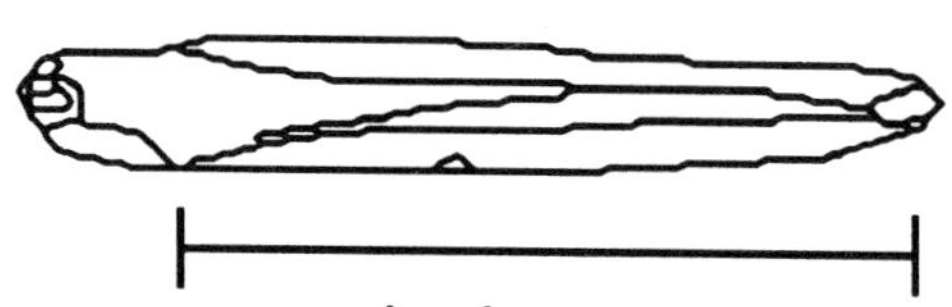

endscraper shape = 0.32

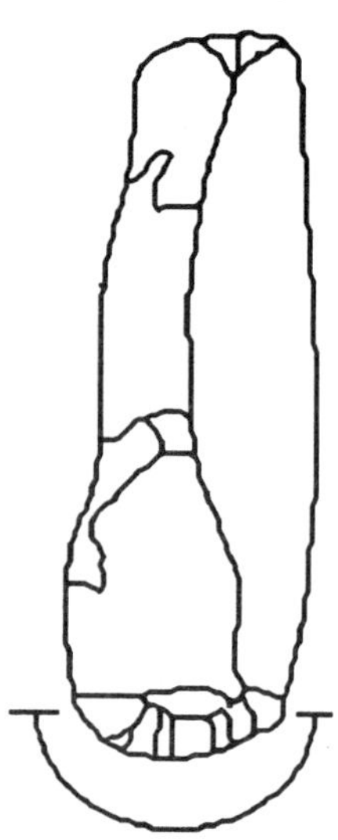

endscraper shape = 0.48

The remaining variables are recorded separately for the dorsal and ventral surfaces. The presence of intentional retouch used to form the tool is most common on the dorsal surface, being initiated from the flatter ventral surface. Retouch can often make the recording of edge wear and polish variables difficult, and sometimes impossible in the case of edge wear. Simply, it is often not possible to differentiate between small flakes detached during intentional retouch and those detached by use and often the dorsal surface is recorded as retouched without further comment.

It is essential to record polish variables both for the ventral and dorsal surfaces as the presence of bifacial polish obviously demonstrates that both surfaces were in contact with the worked material. Also the presence of edge wear on one side and absence on the other is a good guide to whether the motion is uni- or bi-directional.

5.3 EDGE WEAR

The third and fourth groups, macroscopic and microscopic edge wear, describe any modifications to the edge other than that from intentional retouch. These two categories include the **presence, amount and type of fractures**, the **presence and degree of rounding** of the edge, and in the macroscopic category, the **presence or absence of visible gloss** on the working edge.

MACRO-EDGE WEAR

5.3.1 AMOUNT OF FRACTURES

Macro fractures are those seen by eye. The number of fractures is recorded as greater than, or less than **5 fractures per 10mm.** of working edge. This is a simple way of quantifying minor edge wear which may not be significant, and a greater amount of edge wear that may be functionally diagnostic. A value of less than 5 fractures per 10mm. may result from mild accidental damage or be produced by retouch, as even when retouch is unifacial some small flakes may be detached on the opposite surface.

5.3.2 FRACTURE TYPE

Flakes are produced by the normal conchoidal fracture of flint initiated by pressure or percussion against one surface of the edge of a tool, and leave a flake scar that is the negative of the detached flake (Figure 50). **Snap fractures**, appearing as crescentic or half moon shaped fractures, occur when the edge of the tool breaks off under bending stress and leaves no negative scar (Figure 50). **Step scars** terminating abruptly in hinge fractures leave a typical scalar negative scar and are often produced by percussion more directly on the edge, rather than against one side as with flake fractures (Figure 50).

Figure 50

FRACTURE TYPES

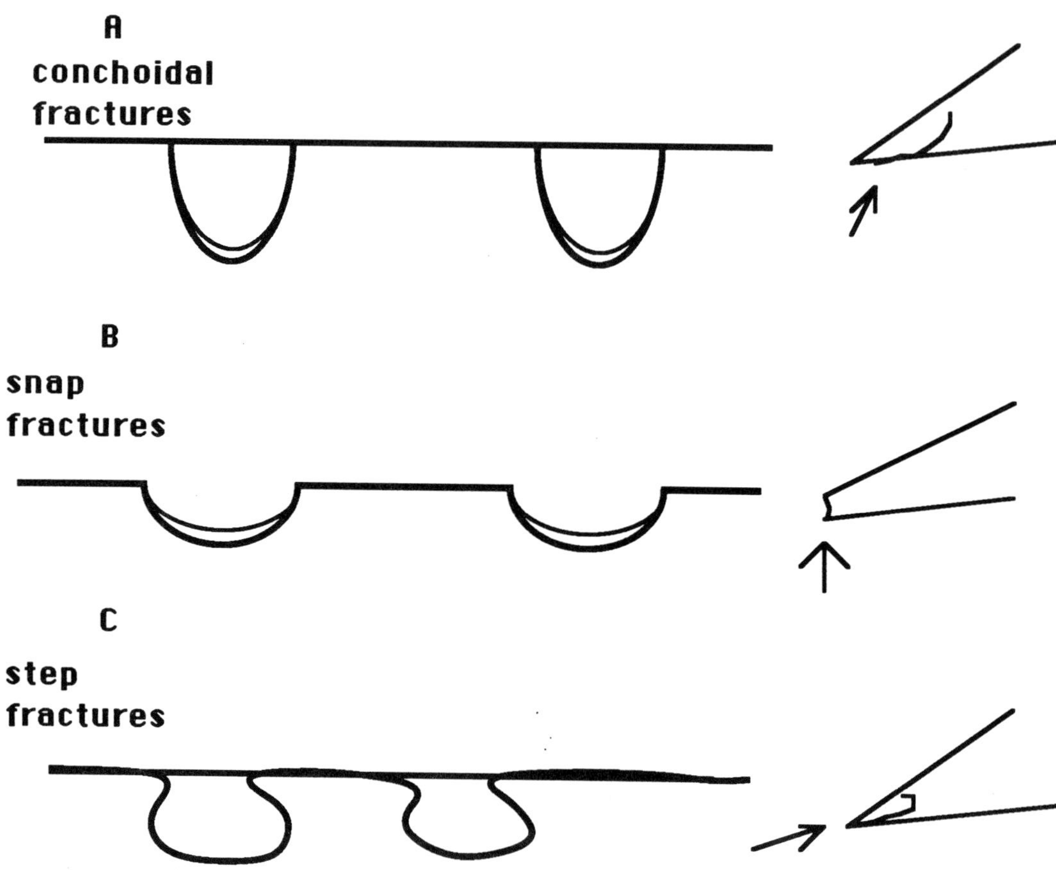

Before recording edge wear, fractures have to be interpreted so as to classify the fractures as **edge wear,** produced by use, or **edge damage** produced by natural processes. Previous attempts to classify fractures have been based primarily on the patterning of the fractures. It has been claimed that there have been "studies that demonstrate a clear difference between damage produced by utilisation and that caused by other factors" (Odell and Odell-Vereeken 1981,90). The main study referred to is that by Tringham et al. (1974), which differentiates wear from damage by saying that damage from various sources produces a random pattern of fractures as opposed to the more regular patterning caused by use. These results are based on mechanistic experimentsof considerable duration, *simulated* experimentswhere tools have been used to carry out a task, rather than being used for a set number of strokes, do not always produce regular patterns of edge wear fractures and sometimes produce no fractures at all. For example test tool number 50 (see SECTION 8) had only macro snap fractures but these were not patterned regularly, being separated from each other at random intervals with few areas of consecutive fractures. Use of a tool on soft materials such as meat often produces no edge wear, or only a few randomly spaced fractures. Even use on a hard material such as bone can produce little or no edge wear if the working edge of the tool is robust like the facet of a burin. When edge wear **is** patterned it may be differentiated from edge damage with fractures produced by such things as soil movement, spontaneous retouch, trampling or *box* damage caused by inadequate post-excavation storage.

Edge wear that does not have consistent patterning with consecutive fractures can be easily confused with edge damage, if the fractures only are considered. The fractures have to be associated with an edge having the functional capability consistent with such fractures. Edge damage usually occurs on any edge of the tool capable of sustaining fractures. The guide is whether the fractures are only present on a potential working edge and absent elsewhere on the tool, an unlikely situation if the fractures were the result of damage. However, such coincidences may occur and corroborative evidence that the tool was utilised is required in the form of the presence of rounding and/or polish.

Recognising edge wear on retouched tools is much more difficult. It is claimed that edge wear can be seen on a retouched edge because it "tends to nick, crush or abrade those part of the larger scars [that is retouch scars] that occur between impact or pressure points (Odell and Odell-Vereeken 1981,96). This kind of micro-fracturing however, can be concomitant with the larger retouch fractures as a result of the same blow; a kind of micro spontaneous retouch. Keeley quotes a case where a Clactonian notch had small flakes scars at the centre of the notch, but these flakes were detached by the same blow that made the notch (Keeley 1980, 27).

Odell, though claiming that retouch can be differentiated from edge wear, admits that it is difficult, as in his blind test he states "All of the incorrectly assessed implements were judged to have been utilised on a substance harder than they actually were. Apparently some of the retouch *was* [authors italics] mistaken for use wear and the resistance of the worked material assessed was thereby exaggerated" (Odell and Odell-Vereeken 1981,118). As mentioned in SECTION 4 , in this method a retouched edge may not be observed and the retouched surface is not assessed. Only fractures on the unretouched (usually ventral) surface are considered.

Other workers have mentioned the various problems in separating wear from damage. Kamminga comments "It is such a delicate and subjective distinction at times that some researchers decline to clearly divide them. There is a good deal of sense in this caution" (Kamminga 1982). Both these authors (Odell and Kamminga) use low power techniques which rely principally on edge wear. Workers using high power techniques tend to diminish the role of edge wear as they concentrate on polish identification.

It is extremely difficult to determine whether or not edge fractures are due to use or damage, especially on retouched edges. The following criteria should be used as a guide.

1) **Patterning of fractures:** When a definite pattern of consecutive fractures occurs on a potentially used edge, this indicates wear. However an unpatterned, random orientated fracture pattern does not necessarily mean that it must have been produced by damage.

2) **Placement of fractures:** If fractures occur on one edge only, and on an edge that is potentially a used edge (determined by edge morphology) but does not occur on other edges of the tool, then this is a strong indication that the fractures are due to use. Edge damage would tend to occur on any edge.

3) **Corroborative evidence:** To assign fractures as definite edge wear, corroborative evidence is required in the form of polish, striations, linear features or rounding.

4) **Retouched edges:** Retouched edges must be treated with extreme caution and often only fractures on the unretouched surface should be considered.

5.3.3 ROUNDING

The presence of rounding of the edge is recorded as **light** or **heavy** which is naturally somewhat subjective, but the two values can be related to experimentally produced examples. Heavy macroscopic rounding can usually be felt by the fingers as a significant smoothing and blunting of the normally sharp edge of the tool.

5.3.4 GLOSS

Gloss refers to the presence of polish that is visible to the naked eye, and is recorded as **present** or **absent**.

5.4 MICRO-EDGE WEAR

The same variables and values are used for micro-edge wear which are recorded by observation through the microscope at **200 magnifications.** The **number of fractures** is recorded as less than, or greater than, **5 per 5mm** as being a more appropriate level of significance at this magnification. The fractures are recorded as fractures visible only through the microscope, excluding large fractures that would have already been recorded as macro-edge wear. Therefore it is possible to have macro-edge wear with micro-edge wear absent. Conversely it is possible to have micro-edge wear without there being any fractures large enough to be recorded macroscopically. One quick guide to recording **micro-rounding** is that when the edge of the tool is horizontal under the microscope a rounded edge can be detected by the difficulty of focussing on the very edge. This is because the rounding produces an edge that cannot be brought within the depth of focus of the microscope at

200 magnifications. Also, if the tool is observed edge on, a rounded edge will appear as a flat area rather than the thin line of an unrounded edge. For an example of heavy micro-rounding see Plate 9.

5.5 MICRO-TOPOGRAPHY OF POLISHED AREA

The micro-topography of the polished area is recorded not as a variable diagnostic of function, but because it may affect the distribution of polish on the natural surface of the flint. For example, if the area is ridged due to the way the flake pulled away from the core, and at 200 magnifications the surface of the flint is seen as a series of ridges and troughs, then the polish tends to be distributed along the ridges. This produces the appearance of a linear polish distribution which may be mistaken as diagnostic of the motion of the tool, whereas it is only a product of the original surface structure of the flint before use (see Section 6.5).

5.6 POLISH DISTRIBUTION

Polish is defined as a visible alteration of the natural surface that increases its reflectivity. This excludes residues which are additive to, rather than an alteration of the surface. Polish distribution is recorded as **continuous** or **intermittent.** Again, the original topography of the tool may be responsible for the distribution and so the polish distribution is recorded merely as continuous or intermittent so that this can be correlated with the topographical features already recorded. (Also see SECTION 6.5).

5.7 POLISH DISTRIBUTION TYPES

Polish distribution types are schematically represented in Figure 51. These distribution types account for most of the variation produced by our experiments using tools in simulated prehistoric tasks on a wide range of materials. The types are simply descriptive of where, and in what arrangement, the polish is distributed on the working edge.

A (away from the edge). Where the polish is predominantly distributed in a band not on the very edge of the tool, but away from it (Figure 51a, also see Plate 7).
B (gapped). Where the polish is distributed in a band away from the edge and on the edge, with an unpolished area in between (Figure 51b).
C (edge only/even). Where the polish is distributed evenly along the working edge (Figure 51c).
D (edge only/asymmetric). Where the polish distribution is along the edge but is asymmetric, in that the polish is more invasive along some portion of the edge (Figure 51d).
E (differential). This distribution is when two different levels of polish development are present on the same edge (Figure 51e).

Figure 51

POLISH DISTRIBUTION TYPES

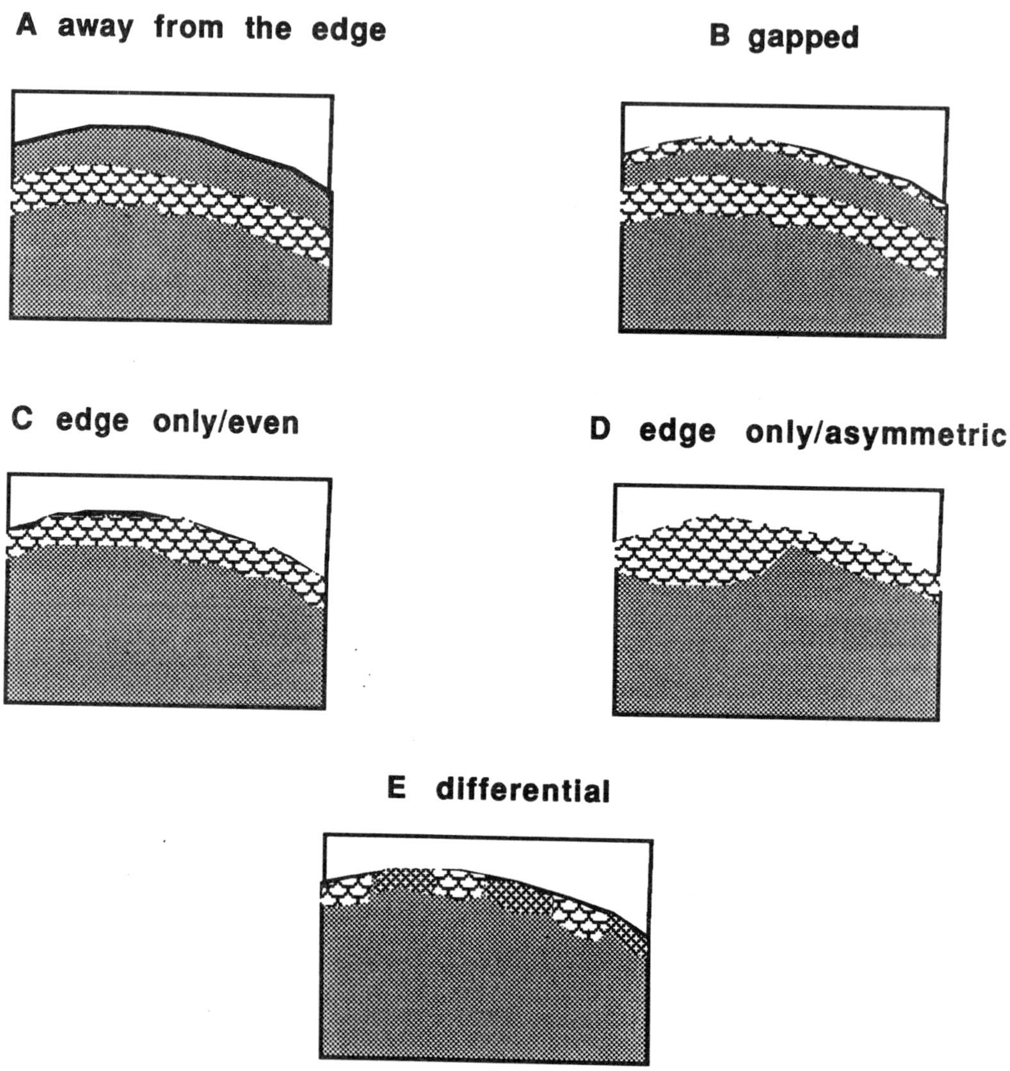

A working edge may have a combination of polish distribution types, for example see Tool 34 (SECTION 6.6 and Figure 60).

5.8 INVASIVENESS

Invasiveness is a measure of how far the polish extends away from the edge. The measurement is recorded as less than or greater than half a diameter of the field of view through the microscope at 200 magnifications (Figure 52). This represents distances of < **100 µm.** (edge only), between **100 µm. and 400 µm.** (<0.5D), and more than **400 µm.** (>0.5D).

Figure 52

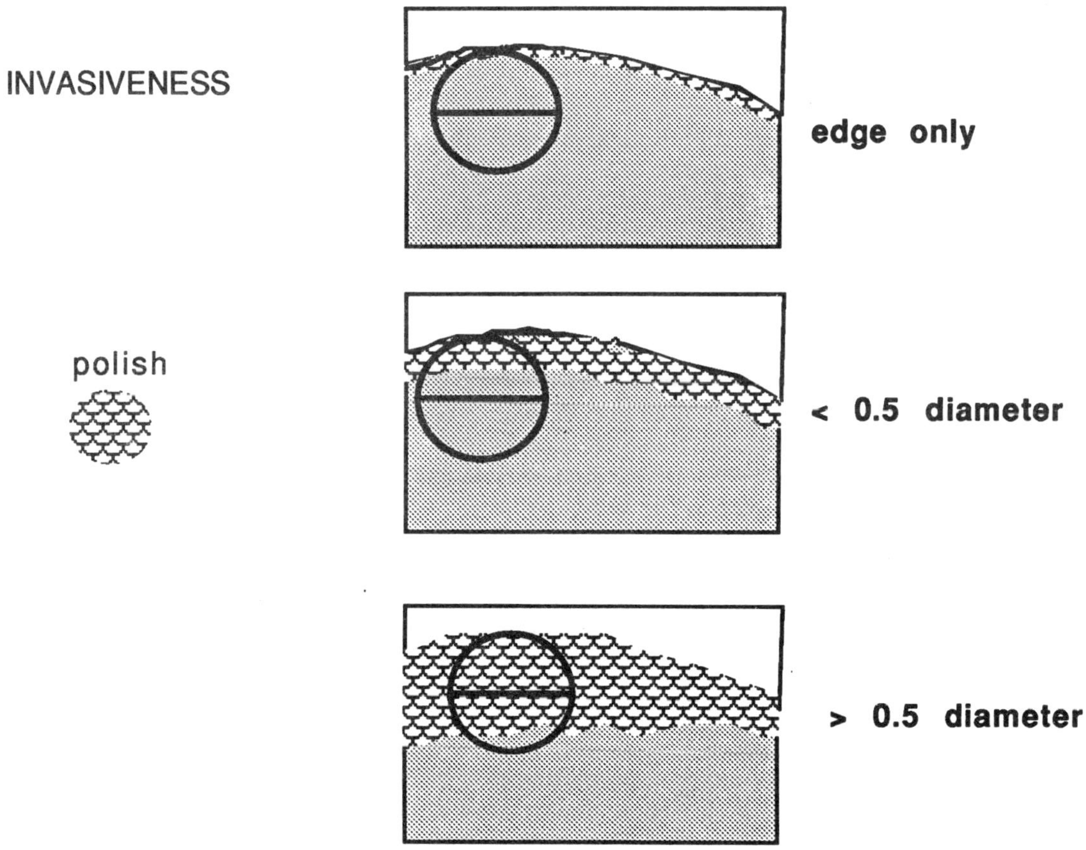

5.9 LINEAR FEATURES

Linear features are defined as lines of polish, and their orientation to the working edge is recorded as parallel, perpendicular or angled (Figure 53 and see Section 4, Figure 37).

Figure 53

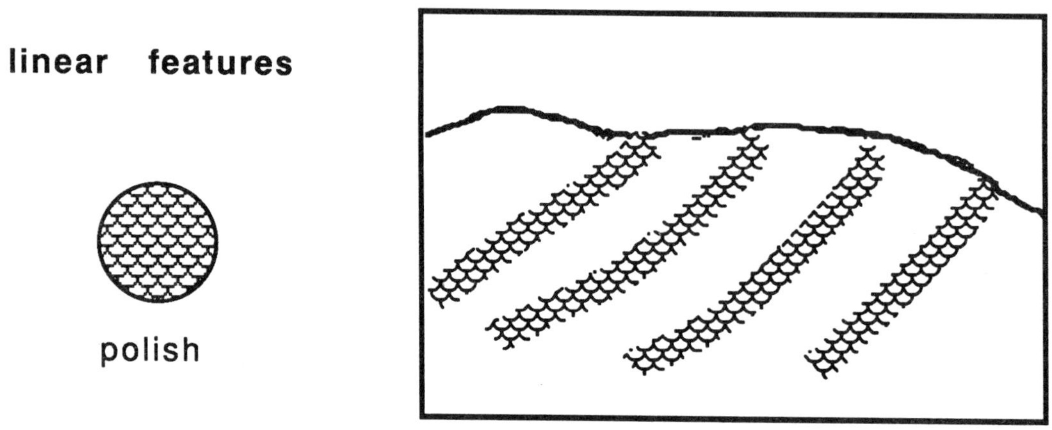

5.10 STRIATIONS

Striations are scratches or grooves in the polish (Figure 54), and are recorded in the same way as linear features.

Figure 54

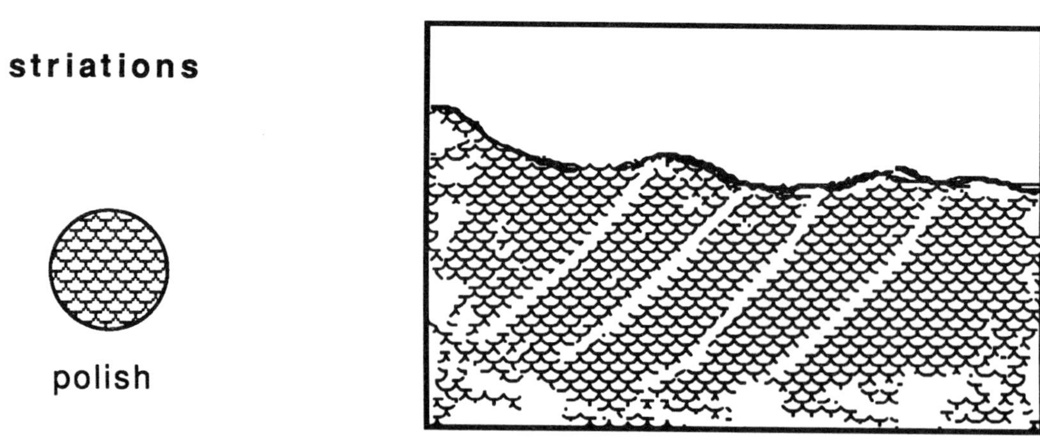

5.11 POLISH DEVELOPMENT

The last category, polish development, is broken down into the schematic representations seen in Figure 55.

Figure 55

POLISH DEVELOPMENT

A individual elements

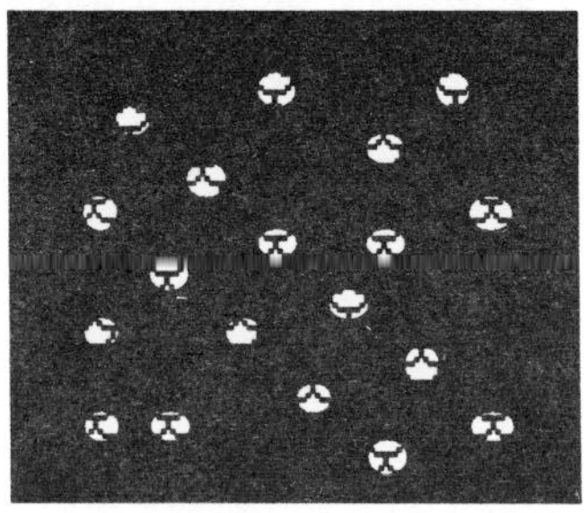

A+

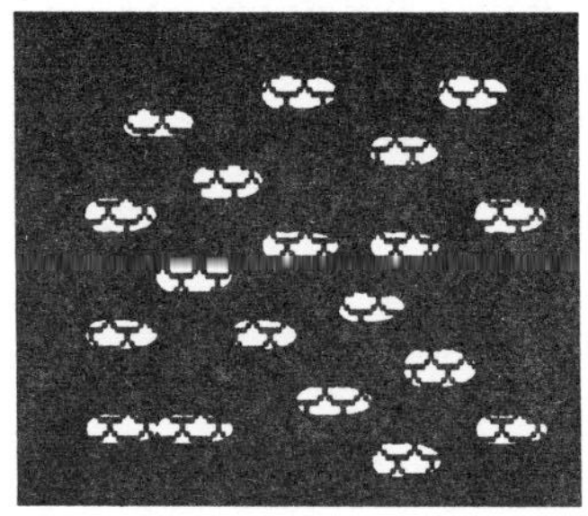

B linked

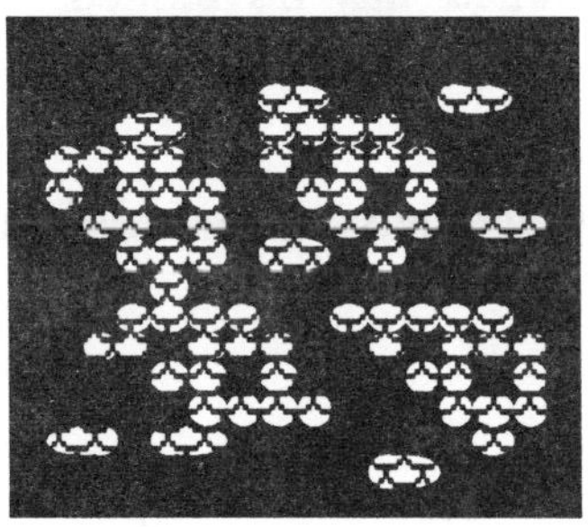

B+

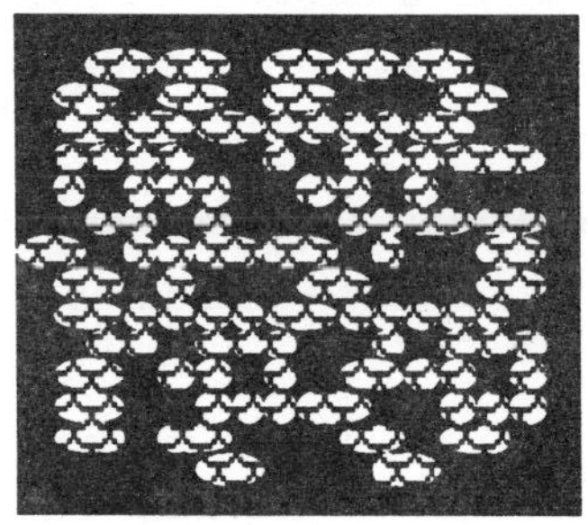

C all over

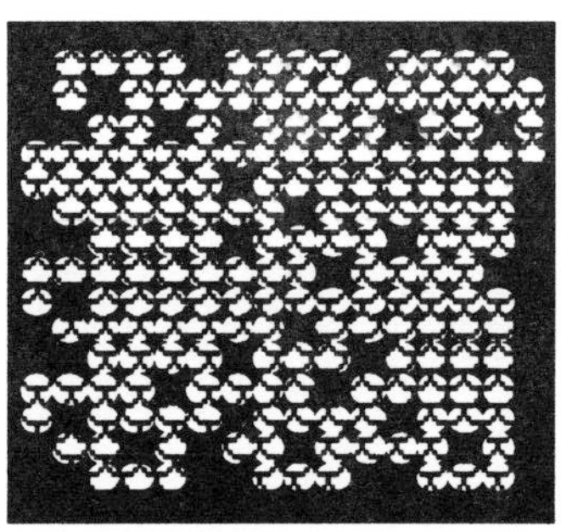

D linear

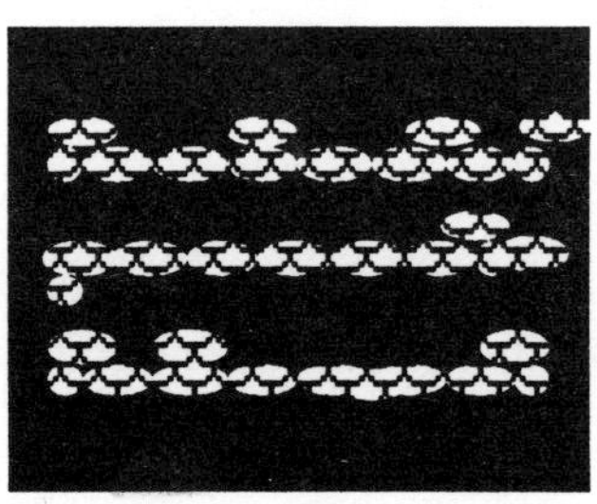

The distribution types represent the most well developed area of polish that is present on the tool. In some cases very small areas of well developed polish are present on an edge, usually associated with a topographical features such as a ridge. These extremes, though representing the most well developed area, are often distinct from the more typical level of polish development, and should be ignored when recording the level of polish development.

A: (individual elements). This distribution is defined as having polished elements that are clearly separated from each other within a matrix of an unpolished surface.

A+: This distribution is when the individual elements are larger, but not yet linked together.

B: (linked). This distribution is where the polish elements are joined together, but the majority of the observed area is unpolished.

B+: This is a linked distribution where the linkage has developed sufficiently so that the polished area is approximately equal to the unpolished area.

C: (all over). This distribution is when the linkage has advanced to the stage when almost all the observed area is polished.

D: (linear). A linear distribution is represented in Figure 55, and is a special case where the polish is distributed in linear areas.

SECTION 6:
FUNCTIONAL SIGNIFICANCE OF THE VARIABLES

Having described the variables and how to record them, it is proposed to describe the functional significance of the variables and therefore why they were chosen. As previously mentioned the raw material variables are not diagnostic of function but have to be recorded because of their potential effect on variables that are functionally diagnostic, such as polish distribution. Raw material here refers to such variables as grain size and topography rather than the differences between flint, chert, quartz, etc.

6.1 EDGE MORPHOLOGY

The main value of a variable such as **edge angle** is that it is functionally diagnostic when correlated with other variables such as edge wear. There is a relationship between edge angle and the amount and nature of edge wear and the hardness of the worked material. For example, when a tool having an acute edge angle is used for cutting meat it will have little or no edge wear, whereas the same edge used for cutting antler will have considerable edge wear. This means that the combination of an acute edge angle and no edge wear rules out the tool's use on a hard material, even if the tool was used only for a short period as edge wear will occur almost immediately.

This relationship between edge angle and the hardness of worked materials allows for probability projections to be made about the likely use of a tool. An acute angled working edge is just not suitable for scraping bone for example, as the resulting edge wear makes the tool in-operable after a short period of use. This allows us to make statements about the functional capability of tools, based on the morphological attributes alone. The **profile** of a tool also limits its functional capability. For example, a concave lateral edge of a tool cannot be used for grooving.

Other morphological attributes are less discriminatory but can indicate the more likely use of a tool. An example of this is the use of **edge length** as an indicator of functional capability. A short edge can be used for sawing wood but a longer edge is more efficient and therefore more likely to have been used in the same way as with modern saws. For cutting logs a rip saw is more efficient than a tenon saw, though it is possible to carry out the task with both tools. **Thickness,** as measured here, reflects the strength of the cross section of the tool in relation to the working edge. Thickness can also limit the likely use of tools. For example, a thin end scraper on a blade is more likely to break from a percussive action such as chopping than a thicker more robust tool.

It is assumed that Palaeolithic man would have been aware of such mechanical properties of stone tools since for the production of such a complexity of tools, especially in the Upper Palaeolithic, a technological knowledge is required that demonstrates an awareness of the mechanical properties of stone, which would not only apply to the making of tools but also to their use.

Shape is the only variable to take into account the overall morphology of the tool as opposed to measurements of a specific working edge. The shape measurement is taken using the working edge as the base and reflects the *handle* of the tool. The *handling* properties of tools can often indicate a probable motion, as experience of using stone tools allows for the subjective assessment of how a tool may have been held. The quantification of shape has meant that such subjective assessments can be tested by mathematically correlating the shape measurement with the other attributes to see if tools having similar functions cluster according to shape. An obvious example of how shape can reflect function is in the case of projectile points which are measured differently as indicated. With projectile points the morphology of the piece is integral to its function. (see SECTION 5.2.5.)

6.2 MACRO-EDGE WEAR

The presence of **fractures** obviously helps to interpret which edges of a tool may have been utilised. The number of fractures can often indicate the hardness of the worked material, depending on the susceptibility of the edge to wear. This is partially determined by such attributes as edge angle. For example, scraping a hard material with thin , acute angled edges will produce considerably more edge wear than scraping with a thick, obtuse angled edge. This means that tools like robust endscrapers will not necessarily produce significant edge wear even on the hardest of materials, so that the absence of edge wear does not necessarily mean that the piece was not used. Especially as there may be polish visible microscopically, without any edge wear being present.

Fracture types can help to interpret the motion of the tool. For example, **flake fractures** produced on one side of the edge can indicate uni-directional movement as the fractures are initiated from one side only.

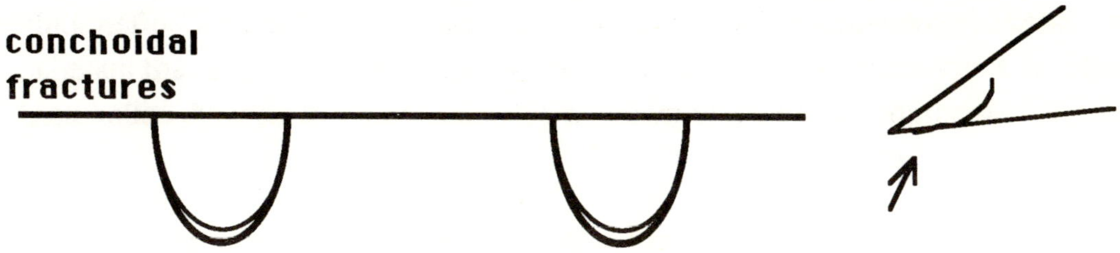

Snap fractures, which only occur on relatively thin edges, can indicate a longitudinal motion as they often occur when cutting or sawing with an unretouched edge, because lateral movement of the tool in the cut produces bending stress on the edge of the tool resulting in the snap fractures.

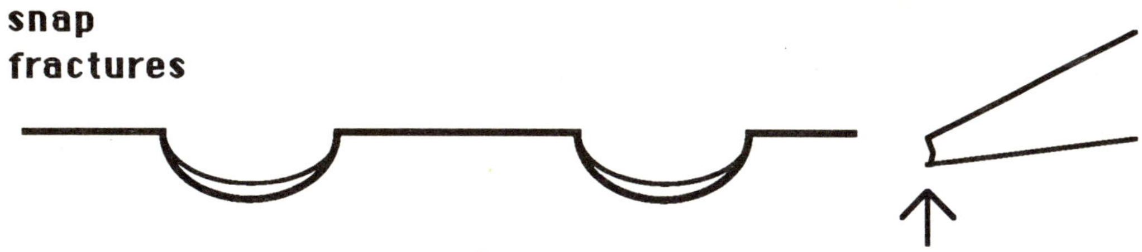

Step fractures are often associated with a percussive motion on the edge of a tool such as with adzing or chopping.

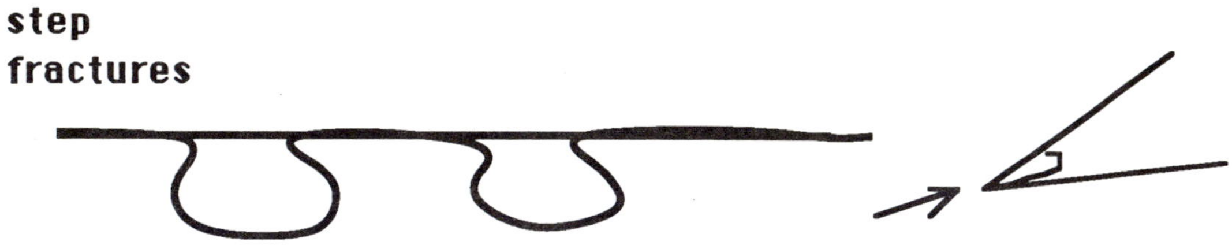

Gloss, that is polish visible to the naked eye, obviously helps to isolate the worked edge, and indicates use on either a hard material that produces a well developed polish quickly, or a softer material used for a considerable length of time, such as with *sickle gloss*. This relationship between hardness of worked material and duration of use is dealt with below (SECTION 6.9).

6.3 MICRO-EDGE WEAR

The features recorded microscopically have the same functional significance as with macro-edge wear. However, those features obviously become more relevant when they are only observed as micro-edge wear, having been recorded as absent macroscopically. For example, an obtuse angled edge such as a burin facet used on a hard material such as bone, may not produce macro-fractures because the strong thick edge is sufficiently resistant to the pressure exerted even with bone, so that the fractures do not occur macroscopically, but very small fractures, often only observable at a minimum of 100 magnifications, do occur and indicate a hard material. This criterion was important for discriminating between bone and soaked antler used on similar edges in blind tests (see SECTION 8.2).

Edge rounding as well as fractures helps in the recognition of worked edges; rounding can also indicate the motion of the tool when the presence of rounding on an endscraper indicates a transverse *scraping* motion for example. It has been suggested by Schutt (1982), that rounding can be separated into uni-lateral and bi-lateral rounding, so that when rounding occurs on one surface this demonstrates a uni-directional motion, and when rounding occurs on both sides a bi-directional motion is inferred. Vaughan (1985, 26) mentions this and points out that when a tool is used in a transverse motion rounding is greater on the surface of the edge in contact with the material, but if the contact angle is high i.e. approaching 90 degrees, the rounding can be equal on both sides. Longitudinal motions tend to produce equal rounding on both surfaces of the edge, so rather than being an indicator of motion by itself, rounding is useful only when considered with other variables.

Rounding can help to eliminate possible worked materials. For example, a longitudinal motion on a soft to medium material, such as sawing fresh wood will (at most) produce only light rounding, whereas heavy rounding would indicate a harder material such as bone. Rounding has been seen as one of the characteristics of use on hide (Keeley 1980). Rounding is produced on such a soft material because the tools which are often used for this activity have thick, obtuse angled edges (e.g. endscrapers), and so the nature of the edge used against a yielding material does not produce sufficient force on the edge to produce significant fracturing, but the abrasion involved can produce rounding. However, this variable cannot be correlated one to one with hide scraping because experiments have demonstrated that rounding sometimes occurs when hide is worked and sometimes not. The amount of rounding may be related to the presence of grit and dirt on the hide rather than the hide itself, and these particles are too small to monitor accurately during experiments.

6.4 MICRO-TOPOGRAPHY OF THE POLISHED AREA

As stated previously, this variable is not functionally diagnostic, but can affect the polish distribution.

6.5 POLISH DISTRIBUTION

The polish distribution is recorded because of the possible correlation with topographical features already recorded. For example, a ridged topography may have an intermittent polish distribution because only the ridges are polished (Figure 56).

Figure 56

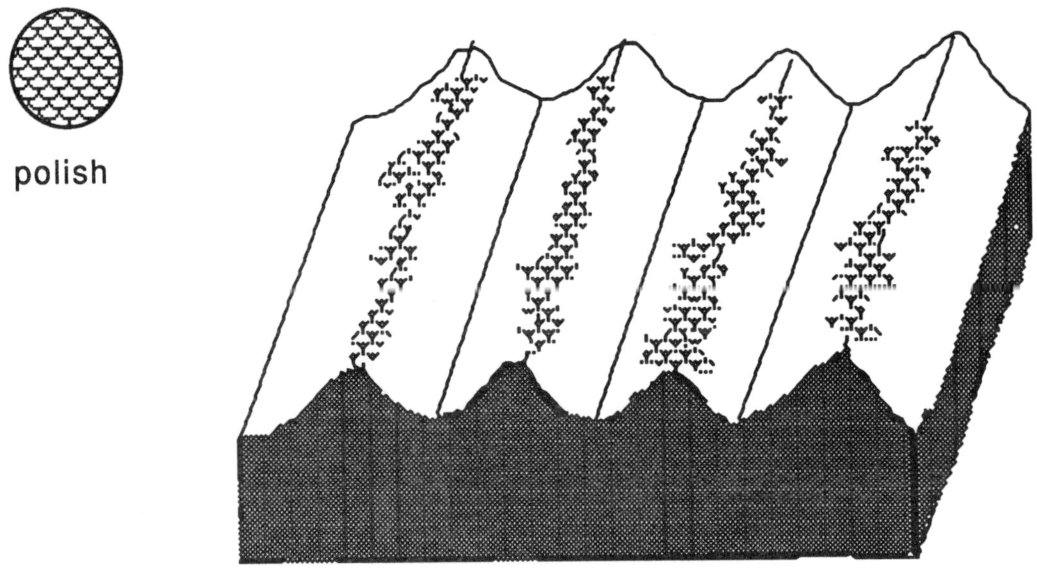

Also if the *valleys* as well as the *ridges* are polished the ridged topography can create the illusion of there being a differential polish distribution (Figure 57 and see SECTION 6.6). The different amount of contact can produce different levels of polish development even though the tool was used on the same material.

Figure 57

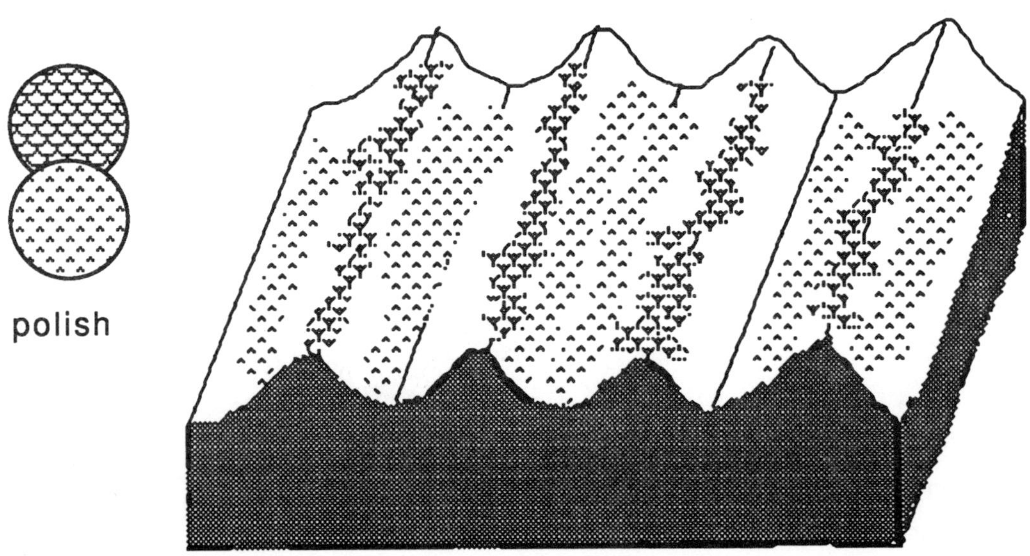

Polish distribution can also indicate possible motions. For example, an edge used transversely at a high contact angle, as in scraping hide, can produce a polish distribution that is **intermittent,** especially with an obtuse angled retouched edge (Figure 58).

Figure 58

transverse motion

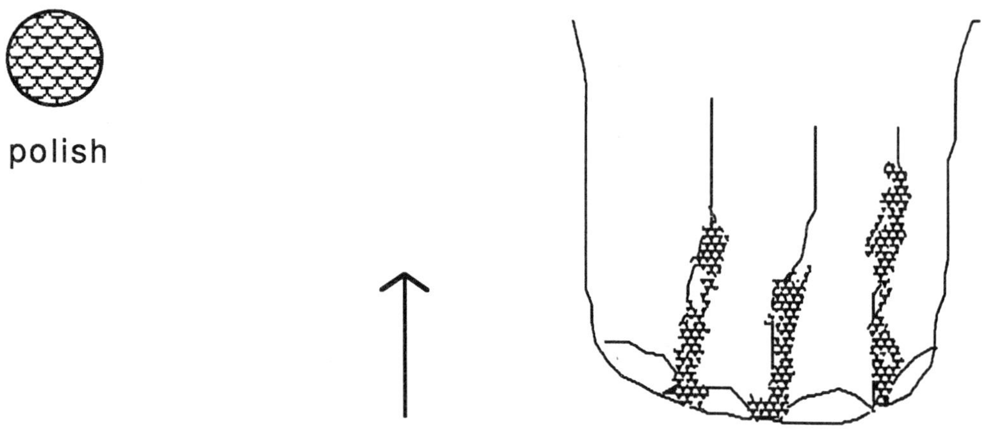

The projections caused by the retouch will often be polished without the intervening surface being effected. If the same edge is used longitudinally a more **continuous distribution** would be expected (Figure 59).

Figure 59

longitudinal motion

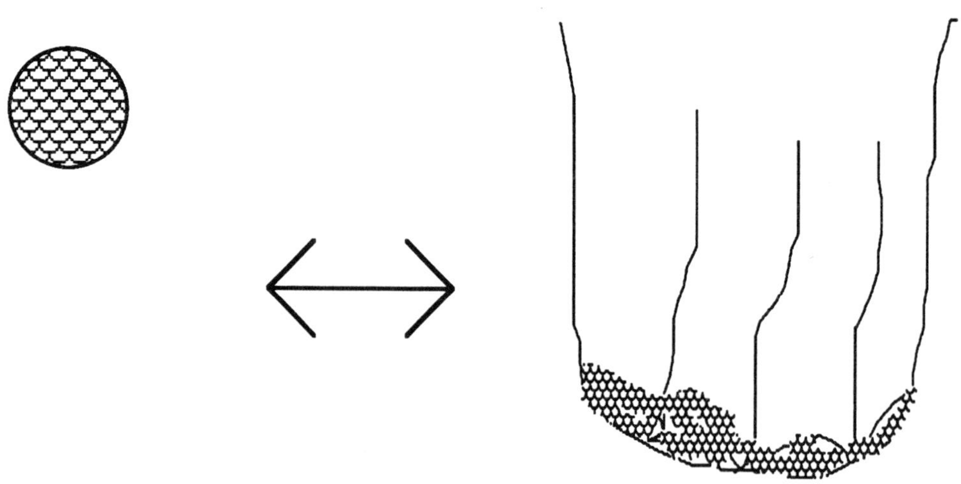

An intermittent or continuous distribution by itself is not very informative. It is only when such variables are considered in relation to others that the origin of the distribution can be interpreted and related to function.

6.6 POLISH DISTRIBUTION TYPES

Type A (**away from the edge**), distribution indicates that the edge is not in direct contact with the worked material, i.e. an edge having type A distribution is not the cutting edge of the tool. This distribution is caused by the surface of the tool rubbing against the worked material. This most often occurs with grooving tools, because the edge is rubbing along the side of the groove rather than cutting into the worked material.

away from the edge

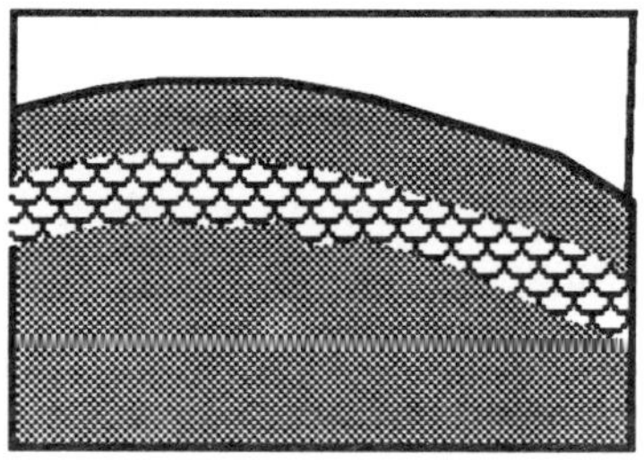

Type B (**gapped**), distribution is when the edge is cutting into the material and the surface is being rubbed against the material. Again mostly associated with grooving, but type B can occur with some cutting motions.

gapped

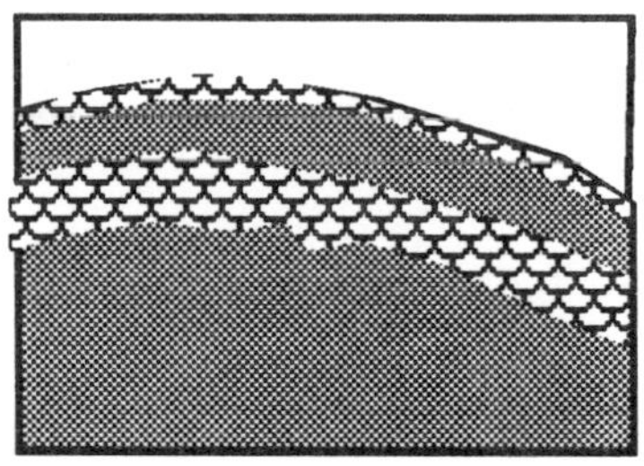

Type C (**edge only/even**), is the most common distribution as being simply an area of polish of equal invasiveness along an edge. This type of distribution can be produced by a number of activities and therefore is not functionally discriminatory.

edge only/even

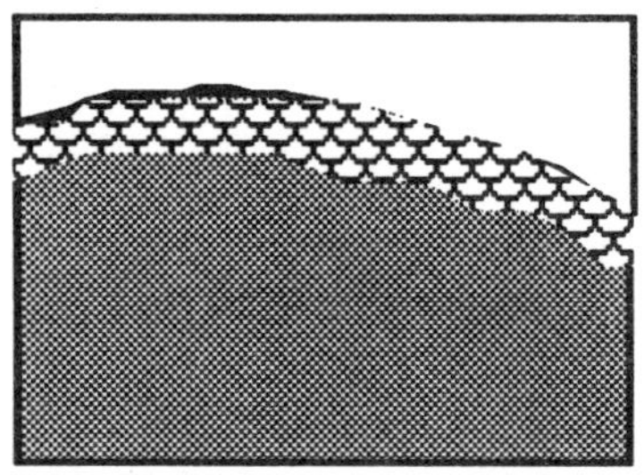

Type D (**edge only/asymmetric**). The asymmetry of this distribution often indicates the leading aspect of the edge and so helps to determine the motion of the tool.

edge only/asymmetric

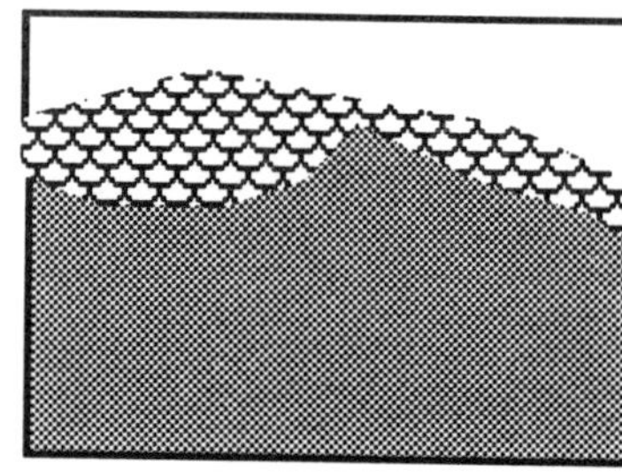

Type E (**differential**), distribution is when at least two different levels of polish development are present on the same edge. This can indicate that either two separate activities are involved, or that two quite different materials are involved in the same activity as when cutting plants on a wooden board. Also topographic features may be responsible for differential polish development as mentioned in SECTION 6.5

differential

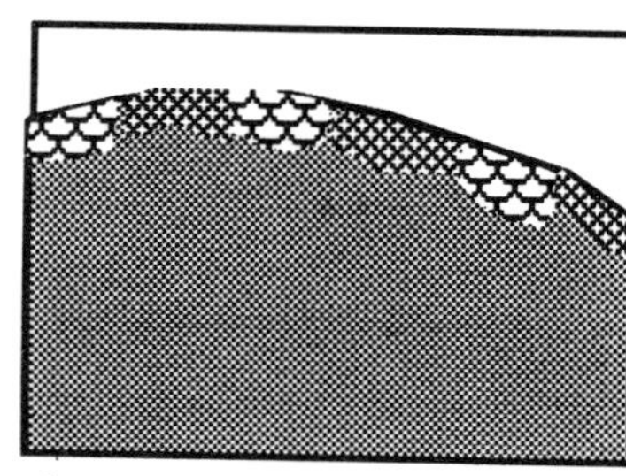

Some tools can have more than one type of distribution. In order to establish the distribution type of test tool 34, the working edge can be drawn as a straight line. Then the distribution is seen to be both *asymmetric* and *away from the edge* (see Figure 60).

Figure 60

tool 34

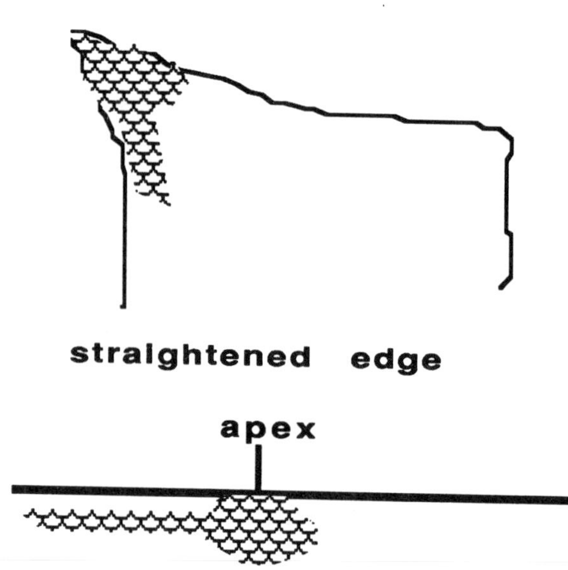

This combination of distribution types is highly diagnostic of the motion of the tool, as it is produced by the point being the leading aspect of the working edge, hence the *asymmetric* distribution. The *away from the edge* distribution is produced by the tool's surface rubbing along the inside of a groove, rather than actually cutting into the material (see Plate 7). The only motion consistent with this combination of distributions is grooving, and this tool was correctly identified as a grooving tool in a blind test (see Figure 82).

6.7 INVASIVENESS

Invasiveness indicates the amount of the tool's surface that was in contact with the worked material. This is most often associated with the amount of penetration of the material, which can indicate whether the material was soft, medium or hard. However, the contact angle can also affect the invasiveness of the polish. For example, a hide scraper used at an acute contact angle will have a more invasive polish that one used at a more obtuse angle (Figure 61).

Figure 61

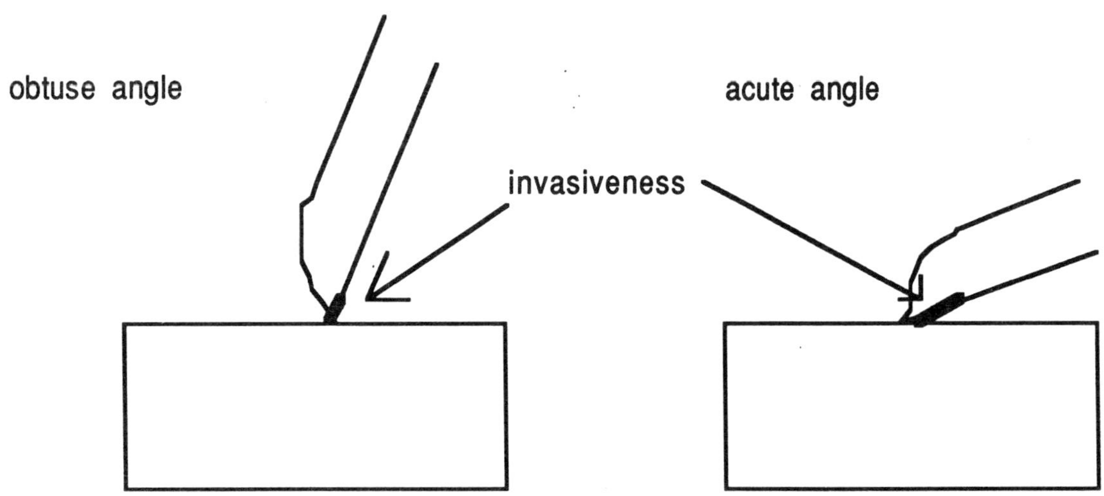

6.8 LINEAR FEATURES AND STRIATIONS

Linear feature is the term used here to describe a line of polish. These are often referred to as striations by other analysts (Vaughan refers to them as *superficial striations* 1985, 24), but a striation by definition is a groove in the material as opposed to a line of polish. The reason for differentiating between linear features and striations is that though the precise mechanism of polish formation is not fully understood, linear features and striations have different mechanical origins. Linear features are polish produced from contact with the worked material, and striations are voids within that polish. Rather than produce a list of types of striations, as others have attempted (Mansur 1982), in this method these linear indications are recorded in two simple categories that are easily defined and therefore easily recognised during observation.

It would also appear that the range of striations is a continuum rather than falling into precise types that can be associated with a particular activity. A simple dichotomy of linear features and striations would seem the most practical way of recording this kind of linear information. The main diagnostic value of both these variables is as indicators of the direction of the motion of the tool, and therefore it is their orientation that is important.

6.9 POLISH DEVELOPMENT

In the method used here we record the polish only according to the schematic representations of Figure 55 (SECTION 5.11). These reflect the amount of polish development without claiming any particular category to be diagnostic of a specific worked material. Other analysts claim that the characteristics of polishes are diagnostic of a specific worked material. For example, the category *linked* is similar to Keeley's *reticular patterning,* associated with wood polish on a rough surface (Keeley 1980, 35). Since the categories represent a continuum of polish development from poor to well developed, almost any category can be produced by almost any material. From experimental evidence we find that the combination of hardness of material and duration of tool use seem to be the crucial variables in the development of polish.

All tools will first develop a polish made up of *individual elements,* whatever material they were used to work, because the highest points of the flint topography are polished first.

individual elements

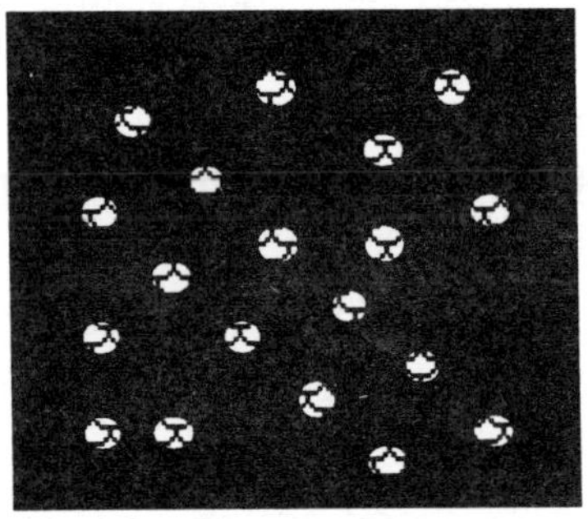

With further use these gradually merge until they become *linked*.

linked

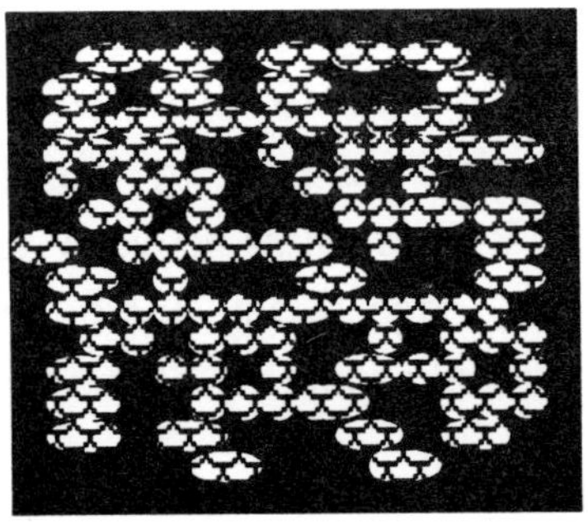

If the tool is used long enough, the polish will be sufficiently developed to become an *all over* distribution.

all over

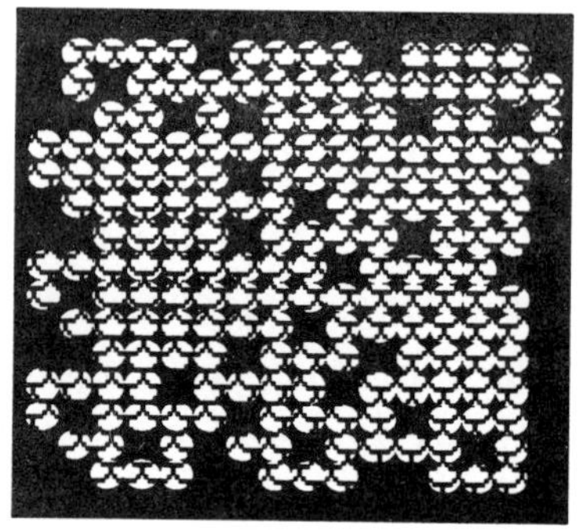

This point can be examined by reference to the admitted difficulty of separating so-called wood and antler polishes (Anderson-Gerfaud 1981 vol 1, 61, Vaughan 1981, 144, Moss 1983, 87, Keeley and Newcomer 1977, 55). Wood will develop a *linked* distribution after a short period of use and then remain as a *linked* distribution for a considerable amount of time. Antler similarly develops a *linked* distribution but then rapidly develops further to an *all over* distribution. Thus, after using tools on wood for 20 to 30 minutes, they will have a *linked* distribution; using tools on antler for 5 to 10 minutes will also produce a *linked* distribution: hence one of the difficulties of separating polishes produced by wood and antler.

This relationship between hardness of material and duration of use in connection with the problems of separating polishes produced by wood and antler is supported by the results of texture analysis (Newcomer et al. 1986, 118 and Figure 4). When tools used on wood and on antler for the same length of time are compared, a difference can be seen, but this is in the form of the greater development of polish produced by contact with antler, because it is the harder of the two materials. When archaeological tools are considered the length of time for which the tool was used is unknown therefore, the level of polish development cannot be taken to be diagnostic of worked material by itself, but can be diagnostic when considered with other variables. Relationships between variables such as the spatial distribution of the polish on the tool, edge wear, and morphological attributes such as edge angle (which limit the functional capability of the tool), can indicate use on materials of a particular hardness.

Following the model of polish development explained in SECTION 1, Figure 34 if polish **X** is observed, then correlations with other variables can eliminate some materials. For example if the working edge having polish **X** had an acute edge angle with little edge wear, then materials as hard as bone or antler would be eliminated. If the motion of the tool has been interpreted as whittling (a motion inappropriate to hide working) then wood would be the logical material to have produced a polish development of **X** on a working edge with an acute edge angle and little edge damage. Interpreting the worked material as wood could then lead to an indication of relative tool use duration from the model presented in SECTION 1, Figure 34.

SECTION 7: INTERPRETATION

The approach to use-wear analysis employed here does not rely on any single variable being diagnostic of tool use, but depends instead on the cumulative evidence from all the variables which leads to a logically consistent functional reconstruction. The interpretation is structured hierarchically so that the level of interpretation in which the analyst has confidence can be stated. The evidence from use-wear varies on different tools and the interpretation can only be made to the level that the evidence allows. The interpretation sheet is illustrated in Figure 62.

Figure 62

INTERPRETATIONS

TOOL NO.
USED AREA:

DIRECTION OF USE: (longitudinal/transverse/rotational)

MOTION OF USE: (e.g. cutting/sawing)

HARDNESS OF WORKED MATERIAL: (soft/medium/hard)

NAMED WORKED MATERIAL: (e.g. wood or soaked antler)

MOST PROBABLE FUNCTION: (e.g. 'scraping hide')

ADDITIONAL INFORMATION: (working angle, hafted, duration of use etc.)

7.1 USED EDGE

A used edge will normally have some use-wear and polish and be functionally suited to some sort of task which is in keeping with the total design of the tool (see SECTION 6.1).

7.2 DIRECTION OF USE

The direction of use is interpreted by the morphological attributes which can indicate either a transverse, a longitudinal motion or a rotational motion. A transverse motion is one at right angles to the working edge and a longitudinal motion is parallel to the edge. A rotational motion involves either a continuous or backwards and forwards movement. The morphological attributes of an endscraper for example suggest a transverse motion while those of a lateral edge of a blade or flake would suggest a longitudinal motion. Polish distribution also indicates motion, so bifacial polish on a long thin edge with little edge wear indicates a longitudinal motion as both surfaces were in contact with the worked material and the lack of edge damage mitigates against a transverse motion. The orientation of linear features and striations, if present, is an obvious indication of motion.

7.3 MOTION OF USE

Motion of use is inferred through the combinations of variables such as edge morphology, polish distribution, edge wear and linear features or striations, enabling the analyst to eliminate motions until only one is consistent with the observations.

7.4 DEFINITION OF MOTIONS

CUTTING: A uni or bi-directional longitudinal motion with the edge parallel to the direction of use and approximately vertical to the worked material, both surfaces being in contact with the worked material (Figure 63).

Figure 63

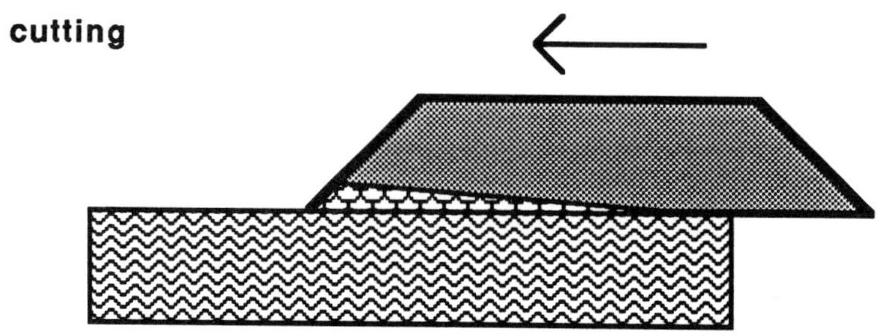

SAWING: A bi-directional longitudinal motion with the edge parallel to to the direction of use and approximately at right angles to the worked material (Figure 64).

Figure 64

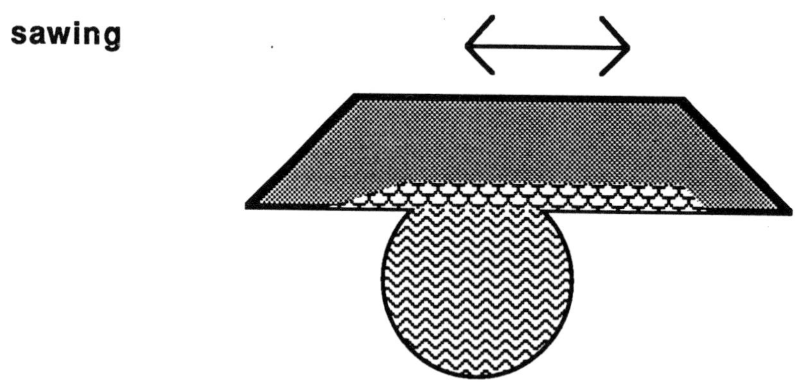

GROOVING: Insertion of the tool into the worked material to create a groove. It may be a uni-directional (Figure 65a) or bi-directional motion Figure 65b), and can be longitudinal or transverse, depending on which aspect of the tool is utilised. The contact angle can vary.

Figure 65

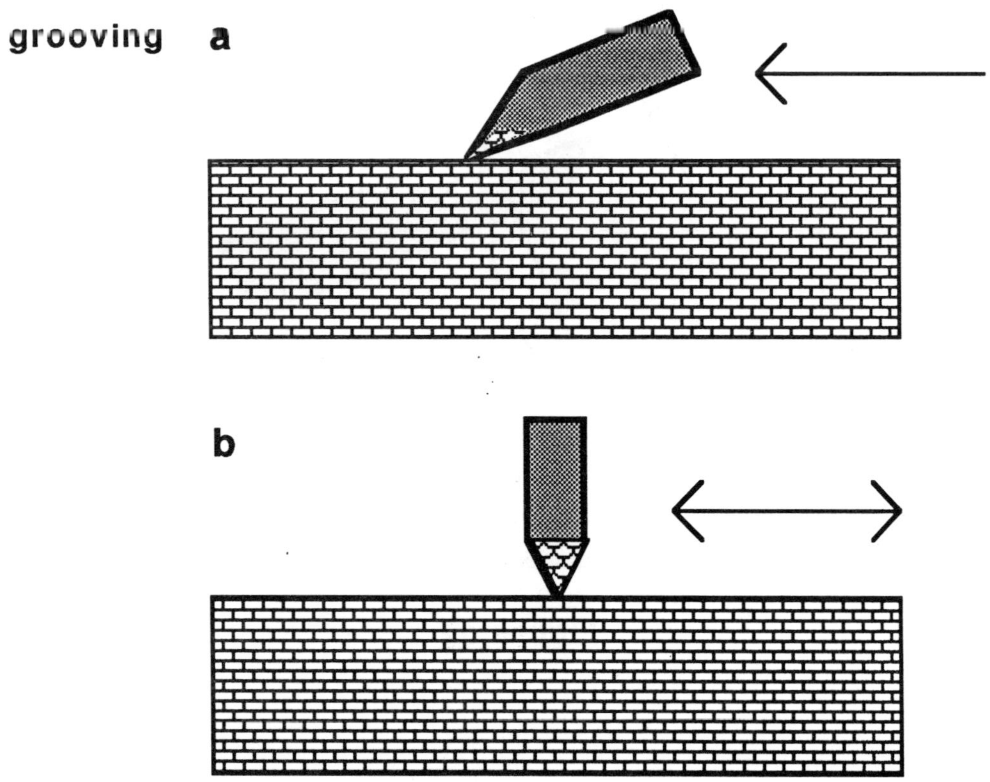

SCRAPING: A transverse action which can be uni- or bi-directional. If uni-directional it can be away or towards the user (Figure 66a). Burin facets used for scraping are also transverse to the worked edge, even though the orientation of the tool is different from that of conventional scrapers (Figure 66b).

Figure 66

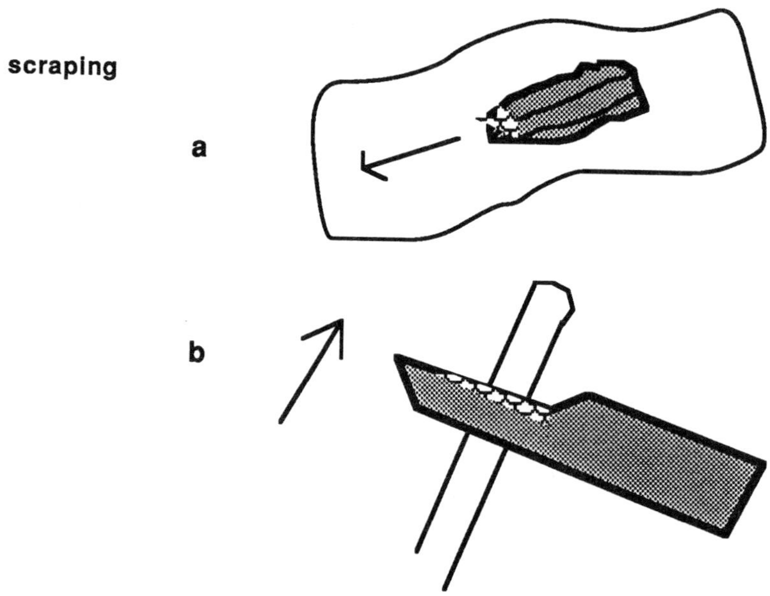

WHITTLING: This is a uni-directional motion often at an angle to the worked material and must involve the removal of slivers of material to differentiate it from the motion of scraping (Figure 67 and see Odell and Odell-Vereeken 1981, 99).

Figure 67

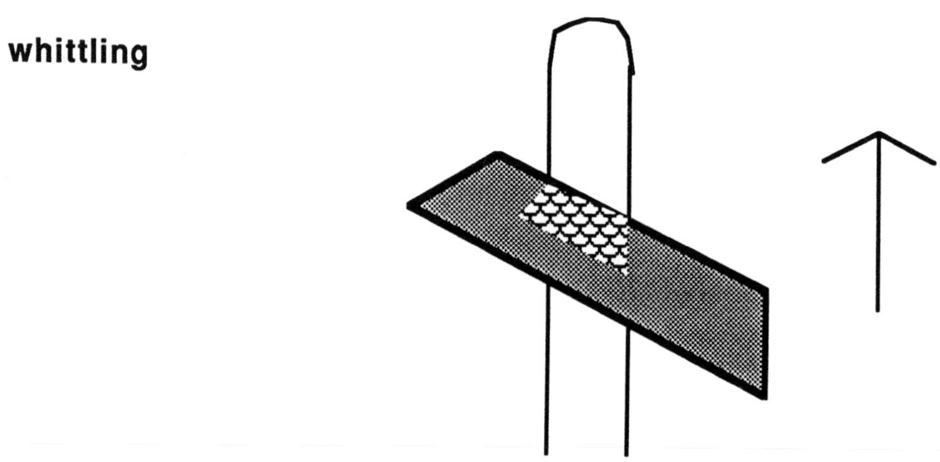

CHOPPING: A percussive motion of use, transverse to the working edge, where both surfaces are in equal contact with the worked material. Contact angle is approximately 90 degrees (Figure 68).

Figure 68

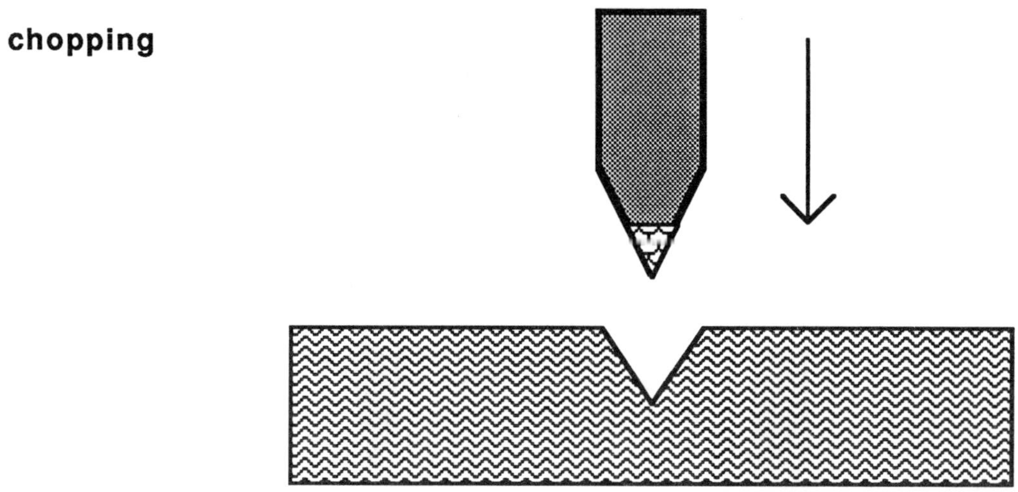

ADZING: A percussive motion, transverse to the worked edge where one surface is in more contact that the other. The motion is designed for *shaping* material rather than for separating it, as in chopping. Contact angle is approximately 45 degrees (Figure 69).

Figure 69

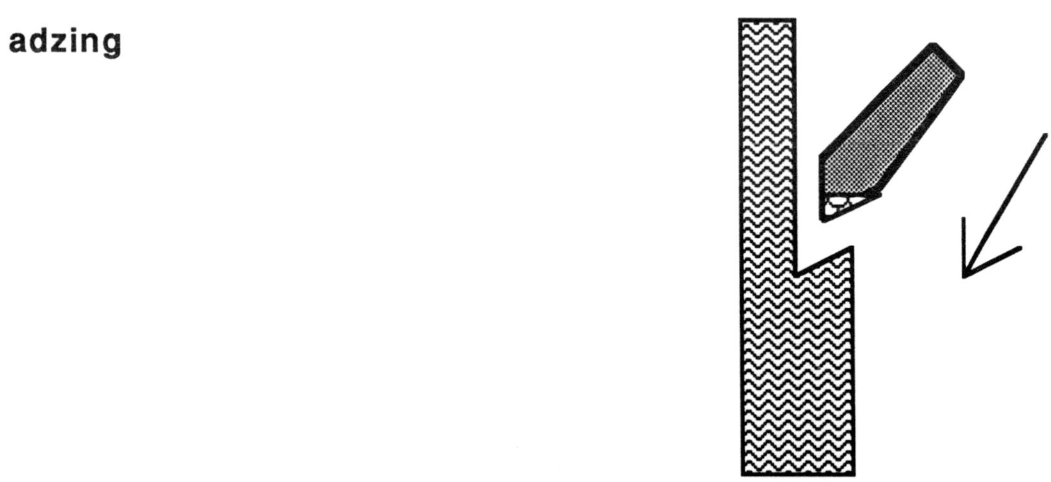

PIERCING: A rotational or transverse motion designed to penetrate material. The motion is transverse when a soft material is penetrated by *pushing* the tool through. For example, piercing hide may involve no rotational motion. (Figure 70).

Figure 70

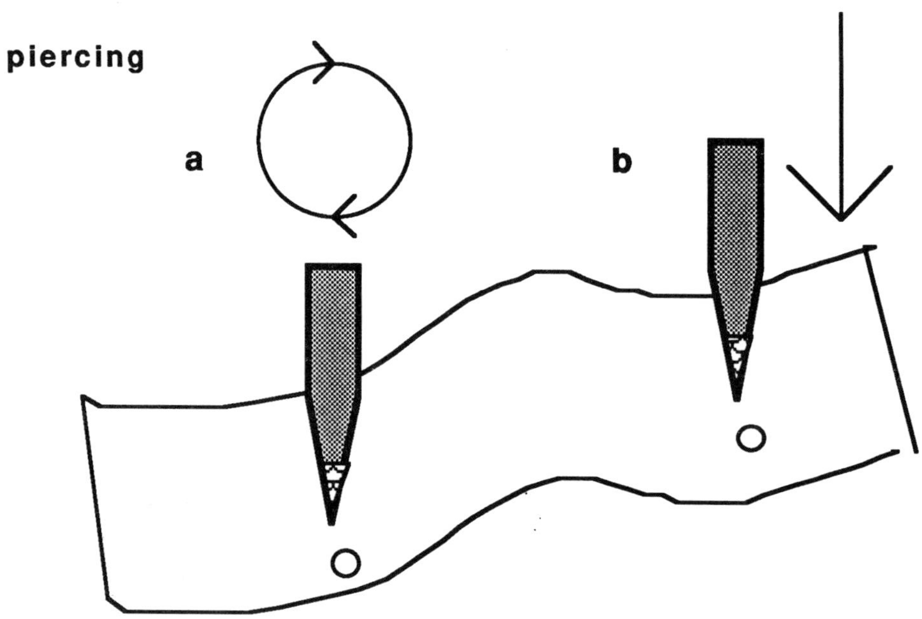

BORING: A rotational motion involving backwards and forward movement designed to penetrate material. Boring can only apply to a resistant material of at least of medium hardness, so that fresh hide, for example cannot be said to have been bored (Figure 71).

Figure 71

boring

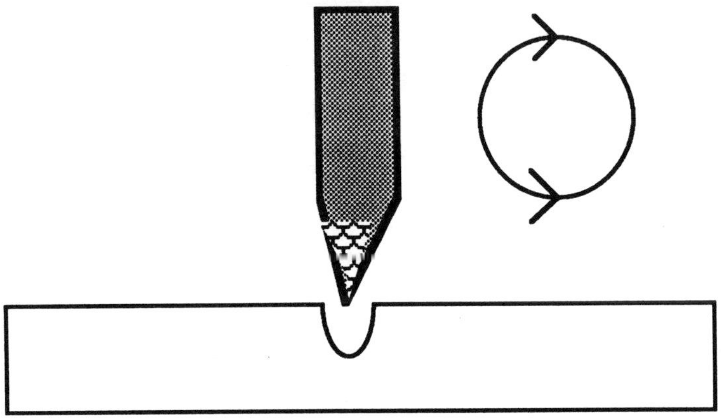

DRILLING: A rotational motion involving backwards and forward movement designed to penetrate material. It is distinguished from boring or piercing since a mechanical device like a bow drill is employed and therefore the tool, by implication, must be hafted, rather than hand held. Though the difference here is essentially of the handle rather than the motion, the use of a mechanical drill produces use-wear that is distinctive from that created by hand held tools (see Chapter 5) and therefore constitutes a different *method* of penetrating material. (Figure 72).

Figure 72

drilling

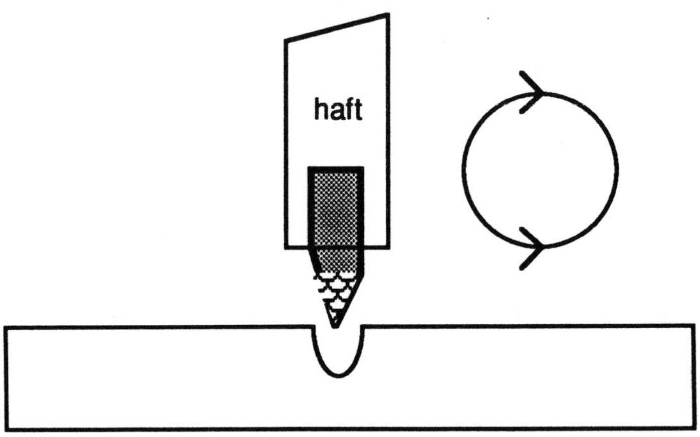

7.5 HARDNESS OF WORKED MATERIAL

Hardness of worked material is divided initially into the categories soft, medium or hard. Though the worked materials form a continuum from soft to hard, they can be classified for interpretative purposes. From extensive experimentation involving hundreds of tools used by a number of people the following classification has been derived. In this context hardness is not the absolute hardness of the material, but instead reflects the *ease of use* of the material with stone tools (for a comparative classification see Odell and Odell-Vereeken 1981,101).

SOFT MATERIALS: meat, plants, woody plants, bark, fresh soft wood, fresh hide.

MEDIUM MATERIALS: other wood, fish, soaked antler, dry hide, soft stone (eg. cortex), horn.

HARD MATERIALS: dry antler, bone, shell, stone. Wood here refers to soft woods such as alder, ash and pine rather than hard woods such as teak and mahogany as these hard woods have not been used in the experimental programme on which this classification is based. These harder woods may well produce edge wear more characteristic of a hard material.

The main variables that can indicate the relative hardness of the worked materials are edge wear, edge morphology and invasiveness, all of which indicate a penetration that is related to the hardness of the material. An inference about the motion of the tool also sets limits to the hardness of the worked material. For example, if the motion of the tool has been interpreted as being grooving or drilling, then at least a medium material must be involved. Cutting in association with a thin edge with little or no edge wear limits the worked material to being soft.

7.6 WORKED MATERIAL

Named worked material: When the motion of the tool has been inferred and the worked material identified as soft, medium or hard, if there is sufficient evidence a more specific interpretation can be made so that the precise worked material can be identified. This is based on the elimination of other materials within the hardness category that has already been established. For example tools 48 and 49 in the blind test (see SECTION 10) were both burins whose facets had been used for scraping a hard material. The polish on these tools was indistinguishable but the presence of micro-edge wear on tool 48 (see SECTION 8.2) indicated that it was used on a harder material than was tool 49. This micro-edge wear was interpreted as being the product of use on bone rather than the softer material of antler. This allowed the worked materials to be identified as bone for tool 48 and antler for tool 49.

The term **most probable function** is used because use-wear analysis is an interpretative technique and it is unsound to make deterministic statements. The evidence from use-wear is rarely, if ever, unequivocal to the point that the precise worked material can be determined to the level of certitude, so that statements of probability based on interpretation are preferable. The same evidence is open to alternative interpretations by another analyst, a situation not uncommon in archaeology generally.

Since the thesis, on which this monograph is based, was written, a computer program has been developed that automates the interpretative process. A description of this program and an example of how it works is given in Appendix 4.

SECTION 8:EXAMPLES

To explain further the process of interpretation a number of experimental tools whose function was interpreted in blind tests will be discussed, demonstrating how the method can lead to precise identifications, and also demonstrating the limitations of functional analysis when there is insufficient use-wear present on the tool.

8.1 TOOL 50

The primary data for test tool number 50 is presented in Figure 74 (also see appendix 3). From these data interpretations are made sequentially to produce a final interpretation of the most probable function of the tool (see Figure 75). The first stage is to identify the used area and this is indicated on an outline drawing of the tool, as in Figure 73.

Figure 73

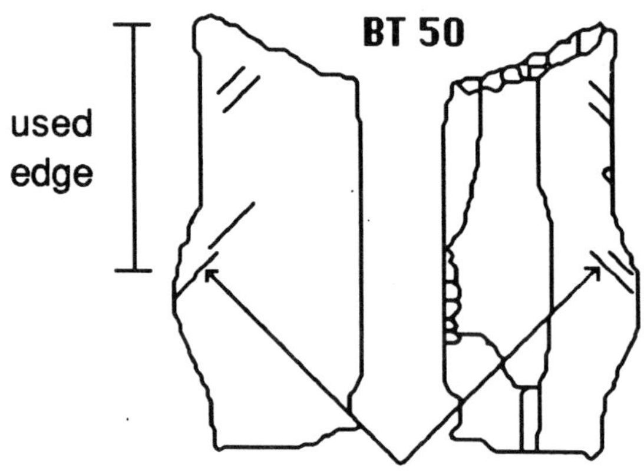

In this case the used area is identified as the **right** lateral edge. This was determined by the morphological attributes used to isolate potentially used edges, and confirmed by the presence of edge wear and polish. The direction of use is first defined as longitudinal or transverse, here transverse based on the suitability of the edge for such a motion, the polish distribution and the presence of angled linear features (see Plate 8). Next the named motion of use is identified by eliminating other possible motions. For example, a scraping motion is eliminated for the following reasons:

1) The edge angle and profile of the tool (**edge angle = 40 degrees, profile = 0.06** i.e. almost straight), made it unsuitable for efficient scraping. If it had been used for scraping far more edge wear would have resulted.

2) The polish is bifacially distributed, though more invasive on the ventral surface, and the
polish on each surface is equally developed. This is not consistent with uni-directional scraping as the leading edge would have a greater level of polish development, nor is it consistent with bi-directional scraping which would not have produced the concentration of polish on one surface.

3) The invasiveness of the linear features indicates penetration of the material to an extent not found with a scraping motion.

The action of whittling is, however, consistent with all these attributes particularly the angled orientation of the linear features. The use of the tool on a hard material is ruled out; whittling a material as hard as bone with an edge angle of 40 degrees, would inevitably produce extensive edge wear that was not present on the tool. The invasiveness and well developed nature of the linear features on the tool argues against a soft material, such as plant or meat, as the yielding nature of such materials does not produce sufficient abrasion to cause such linear features. Use on plants, for example, tends to produce a more invasive polish as opposed to the polish (except for the linear features) on this tool, which is confined to the edge, particularly on the dorsal surface. The motion of whittling also rules out soft materials such as meat. Therefore the worked material must be of medium hardness. The definition of whittling used in this method is that it involves the removal of shavings of material, so that the most probable materials are confined to wood or soaked antler. Other materials of medium hardness, such as scaly fish, are not appropriate to a whittling motion. The penetration indicated by the linear features and the minor edge wear, implies the softer of these materials and therefore wood is the most probable worked material. Thus this method leads to the interpretation that the tool was used for whittling wood, because the combination of this motion of use on this material is the activity most consistent with all the attributes recorded.

Figure 74
OBSERVATIONS

TOOL NO. 50 TYPE truncated blade

EDGE MORPHOLOGY

EDGE ANGLE	40
EDGE LENGTH	45
THICKNESS	5
PROFILE	0.06
SHAPE	2.24

MACRO EDGE WEAR	VENTRAL	DORSAL
FRACTURES	>5 per 10mm.	>5 per 10mm.
FRACTURE TYPE	flakes and snaps	flakes and snaps
ROUNDING	absent	absent
GLOSS	absent	absent
MICRO EDGE WEAR		
FRACTURES	>5 per 10mm.	>5 per 10mm.
FRACTURE TYPE	flakes and snaps	flakes and snaps
ROUNDING	absent	absent
MICROTOPOGRAPHY OF POLISHED AREA	flat	flat
MICRO-POLISH DISTRIBUTION	continuous	continuous
DISTRIBUTION TYPE	edge only even	edge only even
INVASIVENESS	>0.5D	<0.5D
LINEAR FEATURES	angled	angled
STRIATIONS	absent	absent
POLISH DEVELOPMENT	B+	B+

Figure 75

TOOL NO. 50 **INTERPRETATIONS**

USED AREA: (mark on drawing)

right lateral edge-
morphology of tool, backing retouch opposite working edge,
presence of edge damage and polish.

DIRECTION OF USE: (longitudinal/transverse)

transverse:-
orientation of linear features.

MOTION OF USE: (e.g. cutting/sawing)

cutting/whittling-
has a thin edge with edge damage mainly of snap fractures,
the polish is bifacially distributed,
whittling is more probable because of the angled linear features and the polish being more invasive on one side.

HARDNESS OF WORKED MATERIAL: (soft/medium/hard)

medium-
edge wear indicates not a soft material,
to create such linear features requires at least a medium material,
invasiveness of polish indicates penetration of material so a yielding medium material.

NAMED WORKED MATERIAL: (e.g. wood or soaked antler)

wood-
wood is most probable from above evidence, but the lack of polish development possibly indicates a slightly softer material such as a woody plant , like bark.

MOST PROBABLE FUNCTION: (e.g. 'scraping hide')

whittling wood.

ADDITIONAL INFORMATION: (working angle, hafted, duration of use etc.)

hafting is indicated from the linear features on the ventral surface requiring a firm grip, but without covering the flint surface with the hand.
Used at a low angle to produce long linear features with relatively little polish also indicates hafting.
Backing retouch supports this view.

Further insights into the use of the tool can be made. To account for the difference in development of polish on the ventral and dorsal surfaces, a low contact angle (that is the angle at which the tool is used against the material) is required. This is due to the fact that the lower surface in its working attitude had more contact with the worked material and therefore produced more polish than the upper surface. The replication of the motion of this tool demonstrated that it was impossible to hold the tool in the hand at the required angle. If hand held the fingers got in between the worked material and the tool, so that the low working angle could not be achieved. Therefore it was concluded that the tool must have been hafted in the way illustrated in Figure 76 There was no polish produced by hafting but the backing retouch on the left lateral edge of the tool supported this interpretation. The reconstruction of the motion of use, the worked material and the way in which the tool was hafted, matched exactly the way in which the tool was used by the experimenter.

Figure 76

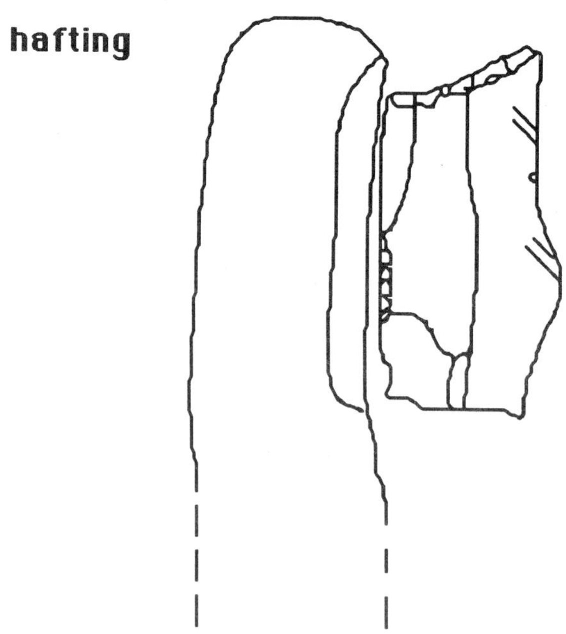

8.2 TOOL 48

The primary data for test tool 48 are presented in Figure 78 (also see appendix 3).

Used area: The used edge was identified from the morphology of the tool (i.e. the burin facet) and by the presence of edge wear and polish (see Figure 77).

Figure 77

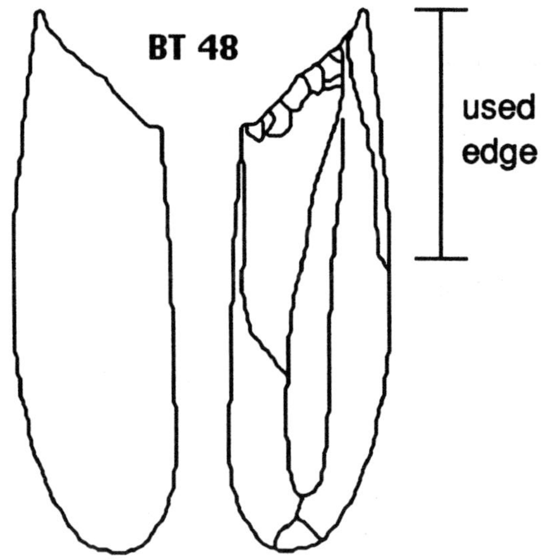

Direction of use: The direction of use was interpreted as being transverse because the polish was along most of the length of the facet and not concentrated on the bit. Also micro-edge wear along the facet indicated a transverse motion as this kind of motion would be required to produce sufficient pressure on the edge to produce edge fracturing on such an obtuse edge (**73 degrees**).
Motion of use: The morphology of the tool limits the potential motions to grooving or scraping as there is no sharp edge for cutting and no point for piercing or boring. The transverse motion rules out grooving which is longitudinal and therefore scraping is the only motion consistent with the observations.
Hardness of worked material: A hard material is indicated because of the morphology of the tool, edge wear and the polish being limited to the edge (**invasiveness = edge only**), indicating a lack of penetration. Having an all over (**type c**) polish level of polish distribution indicates a hard material that will create such a level of polish with minimal contact in a short period of time, (if the tool had been used for some time rounding might be expected).
Named worked material: The most likely worked materials are therefore bone or antler. It was decided that the micro-edge wear (**>5 per 5mm.** on ventral surface) would be produced by the harder of the two materials and so bone was nominated.

Most probable function: The interpretation of most probable function was therefore scraping bone.

Actual use:scraping bone for 6 minutes.

Figure 78
OBSERVATIONS

TOOL NO. 48   TYPE truncation burin

EDGE MORPHOLOGY

EDGE ANGLE	81
EDGE LENGTH	25
THICKNESS	5
PROFILE	0
SHAPE	3.9

MACRO EDGE WEAR	**VENTRAL**	**DORSAL**
FRACTURES	absent	absent
FRACTURE TYPE	absent	absent
ROUNDING	absent	absent
GLOSS	absent	absent
MICRO EDGE WEAR		
FRACTURES	>5 per 10mm.	absent
FRACTURE TYPE	flakes	absent
ROUNDING	absent	absent
MICROTOPOGRAPHY OF POLISHED AREA	flat	flat
MICRO-POLISH DISTRIBUTION	continuous	continuous
DISTRIBUTION TYPE	edge only even	edge only even
INVASIVENESS	edge only	<0.5
LINEAR FEATURES	absent	absent
STRIATIONS	absent	absent
POLISH DEVELOPMENT	C	C

8.3 TOOL 47

The primary data for test tool 47 are presented in Figure 80 (also see appendix 3).

Used area: The used area was identified as the retouched edge (see Figure 79).

Figure 79

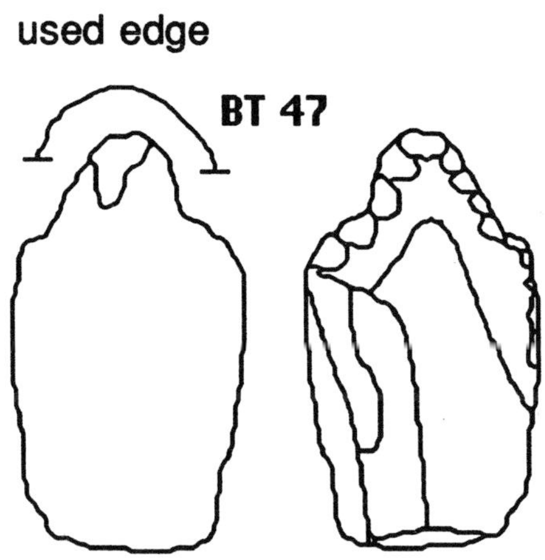

The evidence was the morphology of the tool (created by the retouch) and the presence of polish. There were isolated areas of polish and linear features randomly distributed over both surfaces and some edge damage on the lateral edges. This random distribution and the fractures indicated that they were due to some *post depositional* effect. The possibility that the tool was curated was considered (e.g. by being carried in a pouch or a pocket) but the edge damage required some stress on the edges and so trampling was considered the most likely origin of these non-use traces. Also the polish on the surfaces tended to be of type B in development whereas the polish on the retouched edge was more developed **(B+).**

Direction of use: The morphology of the edge indicated a transverse motion as it was too thick **(10) and too convex (0.8)** to provide a cutting or sawing edge.

Motion of use: Scraping was the only logical motion with the absence of any indications of a percussive motion of use.

Hardness of worked material: A medium material was indicated by the lack of edge wear. A hard material would have tended to create some edge wear and the use of a small area on a hard material would have created a more developed polish (i.e. type c all over), whereas the development of the polish was B+.

Named worked material: Wood or soaked antler was indicated as the harder of the medium materials because of the lack of penetration **(invasiveness = <0.5D)**, but the lack of polish development indicated the softer of these alternatives.

Most probable function: scraping wood.

Actual use: scraping wood for 6 minutes.

Figure 80
OBSERVATIONS

TOOL NO. 47 TYPE shouldered endscraper

EDGE MORPHOLOGY

EDGE ANGLE	52
EDGE LENGTH	30
THICKNESS	10
PROFILE	0.8
SHAPE	0.48

MACRO EDGE WEAR	**VENTRAL**	**DORSAL**
FRACTURES	absent	retouch
FRACTURE TYPE	absent	retouch
ROUNDING	absent	retouch
GLOSS	absent	retouch
MICRO EDGE WEAR		
FRACTURES	absent	retouch
FRACTURE TYPE	absent	retouch
ROUNDING	light	light
MICROTOPOGRAPHY OF POLISHED AREA	flat	retouch
MICRO-POLISH DISTRIBUTION	continuous	retouch
DISTRIBUTION TYPE	edge only even	retouch
INVASIVENESS	<0.5D	retouch
LINEAR FEATURES	absent	retouch
STRIATIONS	absent	retouch
POLISH DEVELOPMENT	B+	retouch

SECTION 9: EXPERIMENTAL REPLICATION

An integral part of this method is the testing of functional interpretations by experimental replication. This can be done by checking the observed wear traces on the tool being analysed against an already existing experimental tool of similar type that had been used in the manner proposed. Alternatively, if such a tool is not part of the extensive collection available at the Institute, then a replica is made and used in the manner interpreted by the analyst. Tool number 44 is an example of this (see Figure 81).

Figure 81

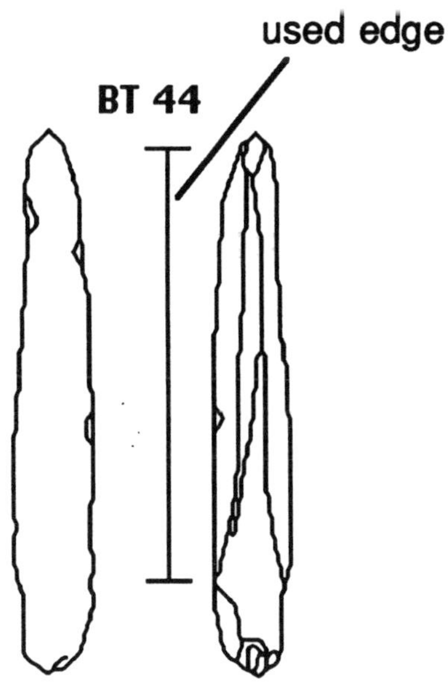

This tool had a thin edge with relatively minor edge damage, but with a well developed polish and heavy rounding (see Plate 9). To create such a polish would require a hard material or a medium material used for a long time. The minor edge damage would suggest a medium material such as wood. The developed polish and rounding on the tool suggests a hard material such as bone. Therefore replicas of the tool were used on wood and bone.

The one used on wood replicated the edge damage, but after 30 minutes the polish development or rounding had not been achieved, and because of the penetration into the wood the polish was already more invasive than on the test tool. Use on bone replicated the polish and rounding but created more severe edge damage than was present on the test tool. Antler was also tried but produced similar results to bone. In order to create the wear traces observed on the tool, it would require a material softer than bone but harder than wood. Therefore attempts to experimentally replicate the wear traces suggested an unknown material.

The term *unknown material* means a material with which the analyst is not experimentally familiar and therefore has no knowledge of any diagnostic wear traces that such a material might produce. In fact the worked material in this case was the cortex of a flint nodule. There was insufficient time to experiment with enough materials to identify this particular one, which accounts for the incorrect answers in the test (Figure 82), though soft stone was mentioned as a possibility. The failure to replicate the wear traces would have meant that if this had been an archaeological tool, then an incorrect identification of either wood or bone would not have been made, and it is better to say *unknown* than to produce inaccurate results.

Figure 82

TEST RESULTS

		USED AREA	MOTION	WORKED MATERIAL
31	Burin bit grooving horn	4	4	0
32	Burin facet scraping antler	4	4	2 (1 partial)
33	Endscraper on hide	4	4	2 (1 partial)
34	Piercer grooving soaked antler	4	4	2 (2 partial)
35	Truncated bladelet cutting green saplings	4	4	4
36	Burin spall boring horn	4	4	0
37	Lunate cutting bark on wood	4	4	4
38	Truncation burin bit pushing holes in bark	3	3	0 (2 partial)
39	Truncation burin facet scraping bone	4	4	2 (1 partial)
40	Piercer piercing hide	4	4	4
41	Endscraper scraping wood	4	4	1 (1 partial)
42	Burin bit grooving shell	4	3 (1 partial)	0
43	Endscraper adzing wood	4	0	4
44	Blade incising cortex	4	4	0
45	Piercer grooving soaked antler	4	3 (1 partial)	3
46	Blade cutting fish	4	4	0 (1 partial)
47	Endscraper scraping wood	4	4	3 (1 partial)
48	Truncation burin facet scraping bone	4	4	3 (1 partial)
49	Truncation burin facet scraping antler	4	3	3 (1 partial)
50	Truncated blade whittling wood	4	4	3 (1 partial)

To be awarded a point for worked material the analysts had to state one precisely correct answer under the heading *most probable function*. If the correct material was among the alternatives mentioned in *named material,* a partial credit was awarded.

SECTION 10: BLIND TESTS

Blind tests are carried out when a group of tools are made and used by an experimenter and then handed over to an analyst who attempts to identify the tasks that the tools were used for. A number of tests have been carried out in order to test the accuracy of use-wear analysis (Keeley and Newcomer 1977) also reported in Keeley 1980, Odell and Odell-Vereeken 1981, Gendel and Pirnay 1982, Newcomer et al. 1986, Unrath et al. 1986, Driskell 1986 and Grace et al. 1988). These tests have been carried out under different conditions and have employed different scoring systems that make it difficult to compare them. All of them, however, have attempted to identify three main aspects of function.

1) The part of the tool that was used **[used area].**
2) The way in which the tool was used **[motion].**
3) The material that the tool was used on **[worked material].**

This scheme follows that set out by Keeley and Newcomer in the first blind test. Three of the above tests had major differences to the others. The Odell test was carried out using low-power techniques concentrating primarily on edge wear. The Gendel-Pirnay test had the precondition that the experimenter would only use worked materials with which the analyst was familiar from his own experiments (Gendel and Pirnay 1982, 251). This in effect limited the range of materials to only 3: hide, wood and bone/antler (no attempt was made to differentiate between the two because the analyst considered that polish produced by bone and antler were indistinguishable). Driskell's account of his blind test does not state which answers he considered correct, half correct or wrong, so that it is difficult to assess the way in which he scored his own interpretations (Driskell 1986).

The remaining tests are more comparable as they did not restrict the experimenter to a certain range of worked materials, so this section will confine itself to a comparison of the Keeley and Newcomer, Newcomer et al. (tools 1-10 only), Unrath et al. and the Grace et al. tests. The Keeley and Newcomer test had only one analyst (Keeley), the remainder were multi-analyst tests having a number of analysts independently assessing the same group of tools. In the Newcomer et al. test there were 5 analysts, and in both the Unrath et al. test and the Grace et al. test there were 4 analysts.

The scoring systems differ in that the conditions of the Newcomer et al. and Grace et al. tests were that a point was scored only if the answers were specifically correct; that is, an answer of bone/antler scored no points. If the tool was used on bone the answer had to be specifically bone to receive a point. It is possible to extract this level of scoring from the Unrath et al. test by using categories A,D and H of their complex scoring system (Unrath et al. 1986,149). Keeley used a system of half points which he added in to his overall score (Keeley 1980,76). The Institute tests do not use this method on the grounds that, for example, answers of **bone/antler** and **possibly wood** are not equivalent to specific answers like **bone** and **wood** respectively. Figure 83 compares Keeley's answers with the correct answers and compares how his test was scored with the way in which the Institute tests are scored.

Figure 83

TOOL	KEELEY'S ANSWER	CORRECT ANSWER	KEELEY SCORE	INSTITUTE SCORE
1	wood	wood	1	1
2	wood	wood	1	1
3	**possibly wood**	**wood**	**1**	**0**
4	**unknown material (possibly vegetable matter or meat)**	**meat**	**0.5**	**0**
5	unused	fat	0	0
6	wood	wood	1	1
7	bone	wood	0	0
8	meat	meat	1	1
9	unused	unused	1	1
10	meat knife (that is a guess, not an interpretation)	hide	0	0
11	**antler (or wood)**	**wood**	**0.5**	**0**
12	wood	frozen meat	0	0
13	**unknown material (but probably vegetable matter)**	**bracken**	**1**	**0**
14a	meat	rabbit skinning	1	1
14b	meat,bone,cartilage.	"	1	1
15	possibly hide, less likely antler	bone	0	0
			total 10/16	total 7/16

The differences are with tools 3,4,11 and 13. Tool 3 "possibly wood" is an indeterminate identification, it could possibly be meat or hide or bone etc. Tool 4: "unknown material (possible vegetable matter or meat)" obviously scores no points. Tool 11: "antler or wood" again is not specific enough to warrant a point. In the Institute tests partially correct results are reported but do not count in the final score. Tool 13: again "unknown material" (however qualified) does not score a point.

The scoring of Keeley's test may be considered harsh, especially the answer "possibly wood" rather than wood, achieving no points. To compare scoring see Figure 87 of the results for the Grace et al. test. Tool 38 used on bark was identified as wood by 2 of the analysts but as they had identified the worked material as being of medium hardness in the **hardness of worked material** category, and bark is considered a soft material (see SECTION 7.5), wood was considered incorrect. Therefore answers of wood achieved no points when the correct material was bark so that the marking of the Keeley test in Figure 83 is no stricter than for the Grace et al. test in Figure 87. This has been done to make the Keeley test comparable with the other tests.

In the tests at the Institute this rigid form of scoring has been adopted from the lessons learned in the Keeley test. "Possibly wood" is an *impossible* answer to judge right or wrong, and saying unknown material and then mentioning two materials and claiming a half point if one of them is correct, leaves it open to analysts to be vague and imprecise and yet to claim positive identifications. This problem of deciphering exactly what an analyst means by vague answers was also a problem in scoring the Unrath et al. test "In attempting to score each analyst's written determinations of used area, motion and contact material of all utilised parts of 21 tools, it was necessary at times to ask an analyst exactly what was meant by an otherwise unclear or incomplete written statement" (Unrath et al. 1986, 151). Some of the answers remained unclear as "after the fact clarification" (ibid,151) was necessary. "The majority of the determinations listed ... are actually those submitted by the participants before they were informed of the real use of each piece" (ibid,151). This means that a minority of the answers were modified after the analysts were told the correct answers !

This vagueness of answers in blind tests is not found so much in reports of microwear analysis of archaeological material. In Keeley's analysis of 46 tools from the Golf Course site, Clacton and 38 tools from Hoxne, only four have question marks and seven are designated as unknown material, this means that 86% (73/84) of his identifications are specific to one material (Keeley 1980). In Vaughan 1985, of 277 worked material identifications 54 have been qualified (eg. bone/antler, unspecified hard material etc.) leaving 223 (80%) that are specific to one material, the majority of which are identified as "gritty dry hide". So Vaughan claims to identify specific worked materials in 80% of cases on archaeological material yet in the blind test in which he took part (Unrath et al. 1986) only 26% was achieved.

A particular case concerning the precision of worked materials concerns the differentiation between bone and antler. As mentioned, Gendel and Pirnay (1982) did not attempt to separate the two claiming that the polishes are indistinguishable. Recent publications (Moss 1987, Bamforth 1987, Hurcombe 1988) have stated that they consider an answer of bone/antler to be correct in blind tests. In Vaughan's analysis of the Cassegros material he used five levels of identification for bone and antler; bone, antler, most likely bone, most likely antler and bone/antler. He claims to identify bone or antler precisely in 61% of cases. In other words, the level of precision as presented in archaeological reports is higher than that to be expected to be achieved in blind tests on experimental tools.

In the Unrath et al. test there were six tools of which 10 edges were used on bone or antler. The answers given are presented in Figure 84.

Figure 84

Answers in **bold** correct by Institute scoring.
Additional answers in *italics* correct in Unrath et al.
A,B, or C represent different edges on the same tool. ml = most likely.

TOOL 4 used on cooked and dried pork leg (meat, fat, sinew, bone)

EDGE	[A]	[B]	[C]
1	bone/antler	**bone and soft animal tissue**	
2	unspecified	antler possibly bone	unspecified
3	*bone or natural traces*	unspecified	*bone or natural traces*
4	unspecified	unspecified	unspecified

TOOL 7 fresh cow bone

EDGE	[A]	[B]
1	bone/antler ml antler	bone/antler
2	antler?	probably hard plant material
3	wood	**bone**
4	antler possibly bone	antler possibly bone

TOOL 7a fresh cow bone

EDGE	[A]	[B]
1	unused	unused
2	unused	unused
3	unused	unused
4	bone/antler ml bone	unspecified

TOOL 13	fallow deer carcass (meat and bone)
EDGE	[A]
1	mammal?
2	*possibly bone*
3	**carcass,bone and soft animal tissue**
4	unused

TOOL 15	soaked caribou antler
EDGE	[A]
1	fresh wood
2	bone/antler
3	soft wood
4	**antler**

TOOL 18	soaked caribou antler
EDGE	[A]
1	*antler (perhaps shell?)*
2	**antler**
3	**antler**
4	unused

Tool four had three edges used on bone, only the "bone with soft animal tissue" would achieve a point in Institute tests; *bone or natural traces* must have been considered correct. Tool 13 had a correct answer (carcass, bone and soft animal tissue) but as two answers were considered correct it is presumed that *possibly bone* was scored correct as *mammal?* seems a little vague. Tool 18 was scored as three correct, and so *antler (possibly shell?)* was considered a specific identification. This means that even accepting the scoring as presented, of 32 possible answers for bone only 5 were correct, and of 8 possible answers for antler 4 were correct. So overall for bone and antler a specific identification was only achieved in 9 out of 40 cases (22%). Under the Institute system the score would be adjusted to 6/40 (15%).

In the Keeley test only one tool was used on bone, which Keeley identified as "possibly hide, less likely antler" (Keeley 1980, 72 Tool 15). In the analysis of the Golf Course site, Clacton one tool was identified as used for "bone boring "and another for "bone graving?". The question mark refers to the *graving* as Keeley states that this tool had a "clear bone polish" (Keeley 1980, 104). Three tools from Hoxne were identified as being used on bone. Thus Keeley identifies five archaeological tools as being used specifically on bone without demonstrating his ability to do this in a blind test. When it was suggested to Keeley that drill bits from Abu Hureyra could have been used on stone rather than wood, he replied that this could not be the case, because "the polish is a highly developed wood polish" (Keeley 1984, 257). There were no reservations about it being possibly wood. So if precision is claimed for archaeological tools then that same precision should be demonstrated on experimental tools.

In the Grace et al. test four tools were used on antler giving a possible score of 16 for the four analysts; 10 out of 16 were antler, at the level of **named worked material** one was bone/antler, three were antler/wood and two were hide. Two tools were used on bone giving a possible score of eight; five answers were bone, two were bone/antler and one was wood/antler. Therefore it was possible to distinguish between use on bone or antler at the rate of 15 out of 24 cases (62%) in this test.

Using the Institute scoring system, Figure 85 illustrates the comparative results of the four blind tests under discussion.

Figure 85

	USED AREA		MOTION		WORKED MATERIAL	
KEELEY	14/16	87%	12/16	75%	7/16	44%
NEWCOMER ET AL.	32/50	64%	18/50	36%	3/50	6%
UNRATH ET AL.	87/120	72%	62/120	52%	31/120	26%
GRACE ET AL.	79/80	99%	72/80	90%	40/80	50%

The Keeley and Grace et al. test results are clearly better that the other two. The aspects of used area and motion are identified to the levels 87% and 75% (Keeley) and 99% and 90% (Grace et al.). This illustrates that the main problem is the identification of the precise worked material, the best result being 50% (Grace et al.). If we look at the scores for the Grace et al. test at the level of **named worked material** (see SECTION 7). Where alternatives are allowed (and incidentally the origin of partially correct results in Figure 82), the score increases to 66%. If the scores are assessed at the level of **hardness of worked material** (equivalent to the group level of Unrath et al.), the score increases to 90%.

Odell has stated the case for carrying out use-wear analysis to different levels, saying that even analysis at the level of hardness of worked material can yield significant archaeological information. "Although accurate assessment of exact material is desired, categories such as 'hard', 'medium' and 'soft' are usually sufficient in answering questions of environmental and human behavioral import" (Odell and Odell-Vereeken 1981,89). Odell claims a success rate of 67% for relative worked material, i.e. separating the worked material into hardness categories. He uses his success rate to argue the case that if such a level of interpretation is required then low power techniques are sufficient. He accepts that if high level information, i.e. specific worked material, is required then low power techniques are insufficient, as he only achieved a success rate of 38% on exact worked material. These figures do not compare favourable with those achieved by Keeley and especially by Grace et al. (67% as opposed to 90% on hardness of worked material, 38% as opposed to 50% on specific worked material). Also, there are problems with Odell's scoring system. For example, on tool number 25 where he did not locate the used edge, (he identified the edge by polar coordinates 4,5 whereas the tool was used at coordinate 8), but his guess of worked material was correct so he awarded himself a point (Odell and Odell-Vereeken 1981, 111 Table 3). In the Institute tests if the correct edge is not located any further identifications are redundant (see Tool 38, Figure 82). It seems strange to claim a point for identifying the worked material on an edge that in fact was unused.

It is particularly informative to compare the Newcomer et al. test with the Grace et al. test as these two tests were carried out under exactly the same conditions and employed the same scoring system. Figure 85 illustrates the improvement in the second Institute test, especially on specific worked material: an increase from 6% to 50%. This was achieved because of the use of the method presented here rather than by using the *Keeley method.* The fundamental difference being that in this method polishes are not regarded as distinctive. Of the 22 variables recorded in this method only 4 are concerned with observations of polish and 3 of these are concerned with polish distribution. The only variable concerned with the visual appearance of the polish is the measure of the level of polish development (see SECTION 6.9). This demonstrates that the method presented here is more accurate than the *Keeley method.* Also, the claim by some microwear analysts that experience is a vital factor in obtaining correct functional interpretations is denied by the fact the most of the analysts taking part in the Grace et al. test were less experienced than those in the Newcomer et al. test. The only analyst that took part in both tests was Grace whose score on specific worked material in the Newcomer et al. test was 1/10 (10%) and improved to 11/20 (55%) in the Grace et al. test. Only a year elapsed between the two tests and in my opinion the method rather than the years extra experience accounts for the difference in results.

The experiments of the Grace et al. test are illustrated in Figure 86, and the results in Figure 87. The results presented are those given in the category **most probable function**(see SECTION 7.6) where only one clear and precise answer had to be given.

Figure 86

EXPERIMENTS

TOOL 31:	Burin bit used for grooving horn for 5 minutes
TOOL 32:	Burin facet used to scrape antler for 5 minutes.
TOOL 33:	Endscraper used to scrape hide for 15 minutes.
TOOL 34:	Piercer used to groove soaked antler for 7 minutes.
TOOL 35:	Truncated bladelet used for cutting fresh green saplings for 18 minutes.
TOOL 36:	Burin spall used for boring horn for 4 minutes.
TOOL 37:	Lunate microlith used for cutting bark into strips on a wooden backing board for 14 minutes.
TOOL 38:	Truncation burin whose bit was used to *push* into bark to make holes.
TOOL 39:	Truncation burin whose facet was used to scrape bone for 7 minutes.
TOOL 40:	Piercer used for piercing hide for 9 minutes.
TOOL 41:	Endscraper on a flake, hafted in ash with bitumen, used for scraping fresh yew branches for 30 minutes.
TOOL 42:	Dihedral burin whose bit was used for grooving shell for 7 minutes.
TOOL 43:	Endscraper used for adzing ash for 20 minutes. This tool was hafted and broke during use.
TOOL 44:	Blade with a natural point used to incise (cutting) cortex on a flint nodule for 11 minutes.
TOOL 45:	Piercer with a broken tip used to groove soaked reindeer antler for 15 minutes.
TOOL 46:	Blade used for gutting and de-finning 2 large trout for 8 minutes .
TOOL 47:	Shouldered endscraper used for scraping ash for 6 minutes and then trampled for 3 minutes.
TOOL 48:	Truncation burin whose facet was used to scrape dry bone for 6 minutes.
TOOL 49:	Truncation burin whose facet was used to scrape soaked reindeer antler for 25-30 minutes.
TOOL 50:	Truncated blade used for whittling pine for 17 minutes. Hafted.

Figure 87

RESULTS

TOOL	1	2	3	4	correct answer	points
31	grooving antler	grooving antler	grooving wood	grooving antler	**grooving horn**	4 0
32	scraping bone	scraping antler	scraping antler	scraping hide	**scraping antler**	4 2
33	scraping hide	scraping hide	scraping wood	scraping stone	**scraping hide**	4 2
34	grooving antler	grooving wood	grooving wood	grooving soaked antler	**grooving soaked antler**	4 2
35	cutting woody plant	cutting plant	cutting plant	cutting woody plant	**cutting fresh saplings**	4 4
36	boring wood	boring wood	boring wood	boring wood	**boring horn**	4 0
37	cutting plant	cutting woody plant	cutting plant	cutting wood	**cutting bark on wood**	4 4
38	piercing wood	piercing bone	piercing wood	unused	**piercing bark**	3 0
39	scraping antler	scraping bone	scraping wood	scraping bone	**scraping bone**	4 2
40	piercing hide	piercing hide	piercing hide	piercing hide	**piercing hide**	4 4
41	scraping hide	scraping wood	scraping hide	scraping hide	**scraping wood**	4 1
42	grooving antler	grooving bone	grooving bone	grooving stone	**grooving shell**	4 0
43	scraping wood	scraping wood	scraping wood	scraping wood	**adzing wood**	0 4
44	cutting woody plant	cutting plant	cutting antler	cutting bone	**cutting cortex on flint**	4 0

45	boring antler	grooving hide	grooving antler	grooving soaked antler	**grooving** **soaked antler**	3 3
46	cutting woody plant	cutting woody plant	cutting dry hide	cutting woody plant	**cutting** **fish**	4 0
47	scraping soaked antler	scraping wood	scraping wood	scraping wood	**scraping** **wood**	4 3
48	scraping bone	scraping antler	scraping bone	scraping bone	**scraping** **bone**	4 3
49	scraping soaked antler	scraping wood	scraping antler	scraping soaked antler	**scraping** **soaked antler**	4 3
50	whittling wood	cutting hide	whittling wood	whittling wood	**whittling** **wood**	3 3

In the group of test tools 31-40, the worked material was not identified on three tools. Two of these tools were used on horn and came into the category of *unknown material* (see SECTION 9). Since the test, experiments have been carried out on horn and it was found to fit between wood and soaked antler in terms of hardness. The fact that all the analysts recognised that these tools were used on a medium material means that they were correct to that level of interpretation. The third tool incorrectly identified, number 38, was a truncation burin used to push holes in bark. This tool had wear traces consisting only of a small area of polish on the bit and this was insufficient to indicate more than the used area with any confidence.

In the test using tools 41-50 a variation was introduced. Five tools were made by one experimenter and five by another. It was considered that the inclusion of a second experimenter would reduce the possibility of interpretations being based on the habits of one particular experimenter, with whose tool use habits the analysts may have become familiar in the course of previous tests. Of tools 41-50, again only three worked materials were not identified. Tool 44 used on cortex has already been mentioned (SECTION 9). Tool 42 was used for grooving shell; all the analysts stated that the tool was used on a hard material, but that there was insufficient information to identify the specific material. Tool 46 was used to cut fish; again the analysts stated that there was not enough information to go beyond identification of the material as of medium hardness, though one did suggest fish as a possibility.

An example of how the use of blind tests can help in the development of analytical techniques is illustrated by the failure to identify the motion of use of test tool 43. All the analysts identified the motion as scraping whereas the actual motion was adzing. A percussive motion of use, like adzing, was indicated by the presence of **step**

flakes (see SECTION 6.2,3) terminating in hinge fractures on the ventral surface. Therefore, the recording of step fractures has been incorporated into the macroscopic and microscopic edge wear categories for future analyses.

The development of these techniques is based on experimental material because it is essential to establish a reliable methodology of use-wear analysis before attempting to deal with the added problems of post-depositional effects on archaeological material. Such post-depositional effects are being introduced into the blind tests. For example tool 47 was trampled for 3 minutes after use and the resulting wear traces were correctly interpreted as being the product of trampling though the analysts were unaware that such post-depositional effects were being introduced by the experimenter (see SECTION 8.3).

Another aspect of this test was that for tools 41-50 after analysing the tools independently and handing in their observation and answer sheets the analysts discussed their findings among themselves prior to being told the actual use of the tools. This was done to see if a consensus interpretation by the 4 analysts comparing notes would produce different results to the independent results. For the test the analysts had to give a precise answer to allow the test to be marked in the way described. For the consensus results the answers were those that the analysts were confident of if these tools had been archaeological specimens.

Tool 41 was confidently identified as being used for scraping and the material was agreed to be medium but could not be specifically identified though the majority said hide. In fact only one analyst was correct, the only example of when the individual analyst proved more accurate that the consensus.

Tool 42 was interpreted as being used for grooving a hard material but it was not possible to identify the precise material, and in fact no analyst could identify it.

Tool 43 was interpreted as being used for scraping wood and that the tool was hafted. The step fractures mentioned above were discussed and the possibility that the tool was used in a percussive motion was suggested especially as the tool had broken during use. The failure to identify the motion in the actual test was a product of assumption, i.e. that an endscraper was most likely to have been used for scraping, though the traces on the tool except for the step fractures were consistent with a scraping motion. The analysts had thought that the step fractures were caused by manufacture. The discussion revealed the evidence that should have enabled the analysts to correctly identify the motion as adzing.

Tool 44 has been discussed in detail in Section 9. The different worked materials: woody plant, plant, antler and bone were a product of guess work and consequently wrong, as all the analysts stated that the traces were unfamiliar and were a product of an *unknown material.*

Tool 45. The consensus was that this tool had been used for grooving soaked antler.

Tool 46. The consensus opinion was that this tool had been used for cutting a soft medium material of unspecifiable type.

Tool 47. The consensus was that this tool had been used for scraping wood and then had been trampled, creating the edge damage and randomly orientated striations .

Tool 48. Consensus was scraping bone.

Tool 49. Consensus was scraping antler.

Tool 50. Consensus was whittling wood .

Therefore, the consensus was that if these tools had been archaeological specimens the following interpretations would have been made with some confidence.

Figure 88

TOOL	CONSENSUS	ACTUAL USE
41	scraping medium material	scraping wood
42	grooving hard material	grooving shell
43	scraping/adzing wood	adzing wood
44	cutting unknown material	cutting cortex on flint
45	grooving soaked antler	grooving soaked antler
46	cutting medium material	cutting fish
47	scraping wood	scraping wood
48	scraping bone	scraping bone
49	scraping antler	scraping antler
50	whittling wood	whittling wood

So if this had been an archaeological analysis, all the motions would have been correct and six materials would have been precisely correct, three would have been correctly identified to the level of hardness and one would have been correctly identified as *unknown*.. The interesting aspect of this consensus analysis is that when the analysts expressed confidence in their precise identifications of worked materials they were correct and, if they did not know the material then they knew that they did not know. This is a very important part of use-wear analysis, to know when the use-wear traces are sufficient to make a precise identification and when they are not. It is just as important to know the limitations of any technique as to extol its virtues. The reason for carrying out blind test at the Institute is not to justify the technique but to test its capabilities.

GLOSSARY

ACTIVITY: Activity is the term used to describe the combination of worked material and motion.
e.g. hide scraping, wood sawing, antler grooving etc.

ATTRIBUTE: Attribute is the term used to name the particular state of a variable.
e.g. a tool has the attribute of a:
40 degree edge angle,
light rounding or
type C (edge only/even) distribution.

CONTACT ANGLE: Contact angle is the angle between the worked material and the working edge of the tool during use.

EDGE DAMAGE: The removal of material from edges by natural processes eg. spontaneous retouch, soil movement, trampling etc.

EDGE WEAR: The removal of material from edges by flaking and/or rounding by use.

FLAKE FRACTURES: Fractures produced by conchoidal fracture.

GLOSS: Polish visible by eye.

LONGITUDINAL MOTION: A motion parallel to the working edge.

MACRO FRACTURES: Fractures that are visible by eye.

MICRO FRACTURES: Fractures that are visible only through the microscope at 100+ magnifications.

MOTION: Motion refers to the way the tool was used.
e.g. scraping, sawing, whittling etc.

PIECE: A piece is a blank i.e. an intentionally struck flake or blade, or a core, that may or may not have been used.

POLISH: Polish is defined as a visible alteration of the natural flint surface that increases its reflectivity. This excludes residues which are additive to, rather than an alteration of the surface.

RETOUCH: The deliberate removal of material from the piece by man.

ROUNDING: Attrition of the working edge to produce a rounded cross section.

SNAP FRACTURES: Fractures produced by bending stress.

STEP FRACTURES: Fractures that are scalar in shape and terminate in a hinge fracture.

TOOL: A utilised piece of stone, whether intentionally retouched or not.

TRANSVERSE MOTION: A motion at right angles to the working edge.

VALUE: The values are scores recorded for the variables.
e.g. edge angle as a number of degrees,
rounding as light, heavy or absent
distribution type as A-E.

VARIABLES: Variables are the individual features that are observed.
e.g. edge angle, rounding, distribution type.

WEAR TRACES: Flaking, rounding or polish produced by use.

WORKED MATERIAL: Worked material is the material that the tool was in contact with during use.
e.g. hide, wood, bone etc.

CHAPTER 5:
DRILL BITS FROM KUMARTEPE

The methodology described here is designed not as a rigid approach to use-wear analysis but more as a framework within which techniques and methods can be adapted to accommodate the particular problems of a specific use-wear analysis. An example of this is the approach to the use-wear analysis of drill bits from Kumartepe, Turkey undertaken for the Dutch Institute of Archaeology in Istanbul. The particular question in this case was whether or not some form of mechanical drilling was involved in the manufacture of carnelian beads. The specialised nature of this study presented some technical problems. First, the evidence of use-wear present on the tips and/or sides of the drill bits meant that a magnification of 200 times was often inappropriate for looking directly on to the tips of the drill bits. The depth of field at 200 magnifications meant that the area was too small to observe the use-wear, 100 magnifications was often more appropriate for observing the polish. This also applied to the observation of the lateral edges of the drill bits because the curvature of the sides rendered the depth of field at 200 magnifications too narrow to assess the length of the concentric striations that were present on the sides of the drill bits, these being a particularly diagnostic feature of mechanical drilling. For looking at the amount of rounding on the tips of the drill bits a magnification of 50 times was often used in order to be able to observe morphological changes to the drill bits that were caused by use. Therefore the analysis of the drill bits did not rigidly follow the observation and recording system presented in Chapter 4. Instead, observation was concentrated on the use-wear features of rounding, presence of polish and circular striation patterns, together with the fractures at the tip caused by rotation (torsion fractures) or by percussion (flute and burin like fractures). In extreme cases rounding of the tips of drills can be observed by eye (see Plate 10). This kind of edge fracturing is especially associated with drilling. Some torsion fractures occur with other rotational actions such as piercing or boring, and fluting fractures are often found on projectile points, but the combination of these two kinds of fractures is almost exclusively associated with drilling. These flute or torsion fractures are special cases and would be recorded as flake fractures or step fractures (if they terminated in hinge fractures as flute fractures, especially burin-like fractures, often do). These would be noted as special cases within the general categories (see Figure 89).

Figure 89

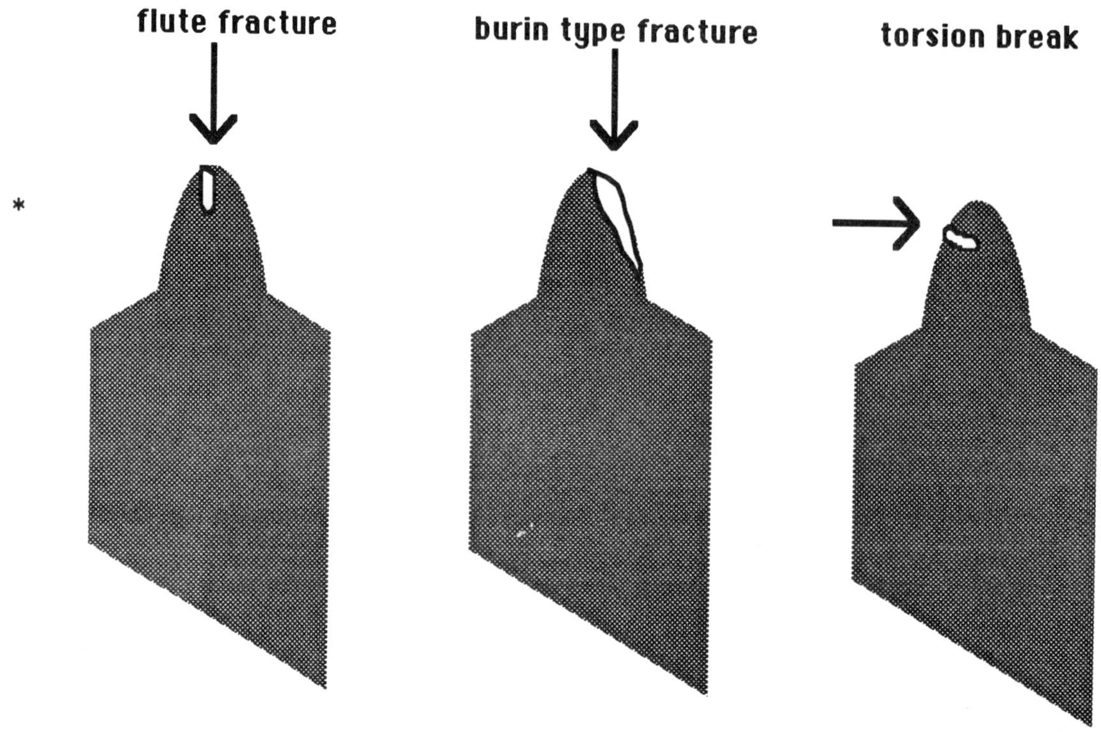

Other morphological features such as edge angle, profile, shape etc. are not discriminatory as this class of tool is almost uniform morphologically within very narrow parameters: mean length = 16.7mm., mean width = 5.7mm., mean thickness = 3.2mm. (see Figure 91, which illustrates the morphological variation of the drill bits from Kumartepe). Indeed, the manufacturing techniques used to produce them were designed to produce this uniformity (Calley and Grace 1988). Another major difference in this study was that not only the tools were examined but also the worked material in the form of partially manufactured or broken beads. This meant that correlations could be made between the wear traces **on** the tools and the wear traces produced **by** the tools.

Kumartepe is a Neolithic site discovered in 1982 by T.J. Wilkinson and G. Stein during a survey of the Samsat area in south-eastern Turkey. This survey was conducted in advance of flooding to be caused by the construction of the Ataturk Dam. It is located on the left bank of the Euphrates. It was excavated in 1983 by a joint team directed by J.J. Roodenberg (Roodenberg et al. 1984). The preliminary report gives a provisional date of the middle to the second half of the 6th millennium.

A bead workshop was discovered in Trench 'A' on the top of a pebble paved surface. A large concentration of flint tools, chips and cores were found. Within this concentration were several thousand micro-borers. The fact that the vast majority of the tools present were of a single type suggested that this area was specifically used to manufacture this kind of tool. The systematic sieving of all the material permitted the recovery of a great number of by-products of manufacture and several broken beads. This prompted the study of the relationship between the micro-borers and the beads. It is notable that nowhere else on the site was such a concentration of bead blanks and micro-borers found.

The use-wear analysis was carried out in two stages. First a sample of the micro-borers were examined and the wear traces recorded and photographed. Then a small experimental programme was carried out to try to replicate the wear traces present on the archaeological material.

It has been demonstrated that carnelian beads can be perforated by a punch technique (Chevalier, Inizan and Tixier 1982), that does not involve drilling at all. So the experimental programme included attempts to replicate this punch technique as well as using a mechanical drilling action. In fact, the preliminary attempts to punch a hole through carnelian by using replicas of the drill bits and striking them with a wooden hammer proved unsuccessful. The drill bit replicas broke almost immediately, within two or three blows. The breakage consisted of either a large burin type break over the entire length of the tool, or the drill bits were completely shattered. The penetration of the carnelian was minimal. When the drill bits were hafted and used with a bow drill the results were more satisfactory. The initial difficulty was in centering the drill bit, because it has a tendency to skid away until a small indentation has been created. Once the drill bit is seated on the surface of the carnelian, drilling can proceed quite smoothly. During the experimental programme it was observed that the more pointed, narrower drill bits tended to break, either in the form of burin breaks at the tip or the whole tip would break away.

The blunter tipped drills tended to work more efficiently and could achieve a penetration of 1mm in 10 minutes. It was further noted that the addition of sand placed in the hole acted as an abrasive and decreased the amount of time necessary to achieve the desired depth of penetration. Microwear analysis of the experimental drill bits (Plate 11a,b) demonstrated that the resulting wear traces bore a remarkable similarity to the wear traces observed on the archaeological specimens (Plate 12a,b).

All of the experimental drill bits exhibited the rounding, polish and circular striation pattern. Often the rounding would form and then break away leaving small areas of polish with striations on the sides of the drill bits, most of the tip having been lost. This was also observed on the archaeological drill bits. Having established the kind of wear traces produced by drilling and by punch techniques, a larger sample of drill bits was analysed. In all 518 drill bits have been examined microscopically, of these 62 have been examined in great detail, that is, every area from the tip to the base on both surfaces has been scrutinised, drawn and all wear traces noted. The remainder have been examined at magnifications of up to 200x, though

100x was usually sufficient to note any wear traces. Observation was concentrated on the tips.

Of the 62 submitted to a full detailed analysis, 38 had either rounding, polish or circular striations, or a combination of these features. This indicates that they were used with a mechanical drill, as the wear traces were the same as those found on the experimental drill bits used with a bow drill. Seven had their tips broken off and so no traces could be observed. The remaining 17 had breaks at the tips that meant only the very tip was intact constituting areas that were too small on which to make reliable observations. Of the 456 examined microscopically, but not in detail, 126 had clear traces of polish, rounding or circular striations, so that of the total sample of 518, 31% had evidence of being used in a mechanical drill. Most of the rest of the drill bits had their tips removed by breakage which was consistent with the breakage patterns seen on the experimental drill bits. Either the tips completely broke off or torsion breaks occurred that removed so much of the tip that polish and striations would not remain. A few drill bits had fluting or burin type breaks probably resulting from use as punches.

Further evidence of the use of a mechanical drill was provided by the microscopic examination of the carnelian bead blanks. 31 partially perforated bead blanks were examined, of these a clear differentiation could be made between those that broke during the initial pecking of the blanks, and those that broke during drilling. 20 of the broken blanks exhibited traces of pecking only. That is, conchoidal fracture scars and an irregular outline to the hole. The depth of the holes ranged from 0.07 to 0.27mm. The seven blanks that broke during drilling had circular holes of regular outline with striations around the rim of the holes as well as in the hole (Plate 13).

The cross sections of the perforations in the blanks were conical and extremely regular. There were three blanks that had been drilled and then abandoned, even though they were unbroken. The bases of the holes of these blanks were observable and matched the size of the tips of the drill bits, and were polished in the same way. The depth of the holes created by drilling ranged from 0.92 to 1.3 mm. One unbroken blank was abandoned after pecking only.

Other evidence for the drill bits being used in a mechanical drill is the presence of a residue on 40 of the 62 drill bits examined in detail. The residue appears to be some kind of mastic and is predominantly on the base end of the bits and possibly represents a residual mastic used for hafting the drill bits. Of 19 piercers examined only 2 had spots of residue on them and no waste flakes had any of the residue. This indicates that the residue is not a natural deposit found on the site, but its existence on the drill bits suggest its presence is due to human activity. The residual material is being analysed and its constituents may clarify whether or not the residue is a material suitable for hafting the drill bits.

These observations of the drill bits and the carnelian bead blanks, allow for the reconstruction of the technique used in perforating the beads. After the blanks were prepared, pecking would have been used to provide a slight indentation in the centre of the blank (the majority of the blanks broke at this stage, i.e. when a percussive action was being used). The indentation would then allow the drill to be placed in position at the centre of the blank without skidding away across the surface. Drilling would then be used to perforate approximately half way through the blank. The completion of the perforation seems to have been achieved by the removal of a conical flake as suggested by Chevalier, Inizan and Tixier (1982, Fig.1). That is, a drill bit is placed in the hole and struck, producing a conical fracture scar, and this completes the perforation. There is one example of a broken blank that was drilled to a depth of 0.92mm. and then broke during the final blow intended to remove the conical flake. No complete beads have been recovered so that we have no information on the final polishing of the beads.

Since this analysis was carried out I have been able to observe some of the beads recovered from Larsa (Chevalier et al. 1982). Unfortunately, none of the micro-borers or *tamponnoir* were available for study. All of the beads observed had perforations whose outlines were irregular as with the *pecked* perforations of the Kumartepe beads. None had circular perforations with concentric striations demonstrating that at Larsa a drilling technique was not used, but the punch technique was used throughout to manufacture the beads. That is, the perforation through half of the bead blank and the final blow to detach the conical flake were all achieved by *pecking*, as opposed to the initial drilling and then final perforation by punch technique as seen at Kumartepe. It is interesting to note that two different techniques were used to achieve the same end with the same material. The other main difference between the sites is that of quantity, there being thousands of drill bits at Kumartepe while only 52 *tamponnoir*, or punches, have been recovered from Larsa. The difference in quantity of material and manufacturing techniques suggest an interpretation of Kumartepe as a *factory* site as opposed to Larsa's *cottage industry*.

Figure 90

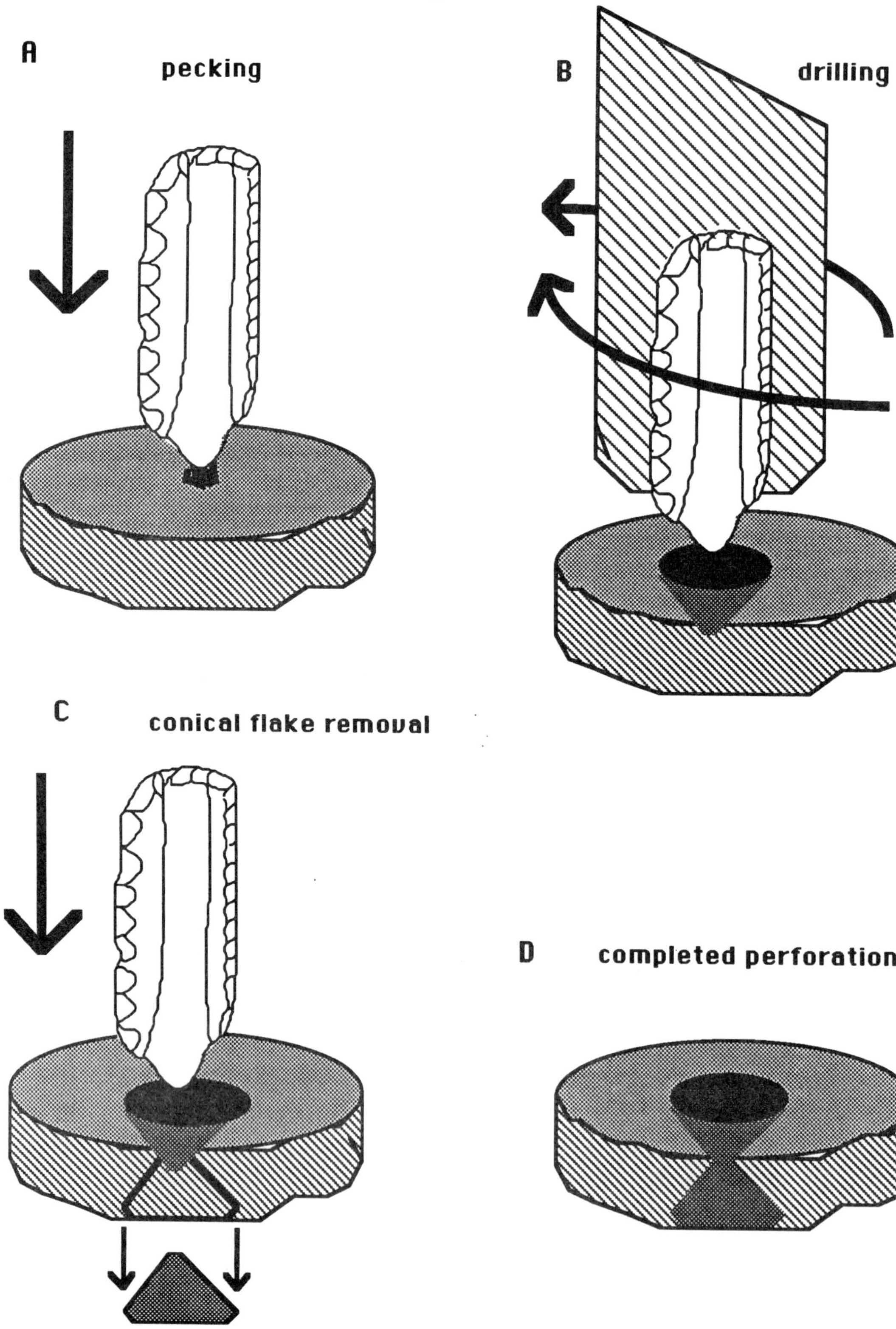
A
pecking
B
drilling
C
conical flake removal
D
completed perforation

Figure 91 SCALE = 3 TIMES ACTUAL SIZE

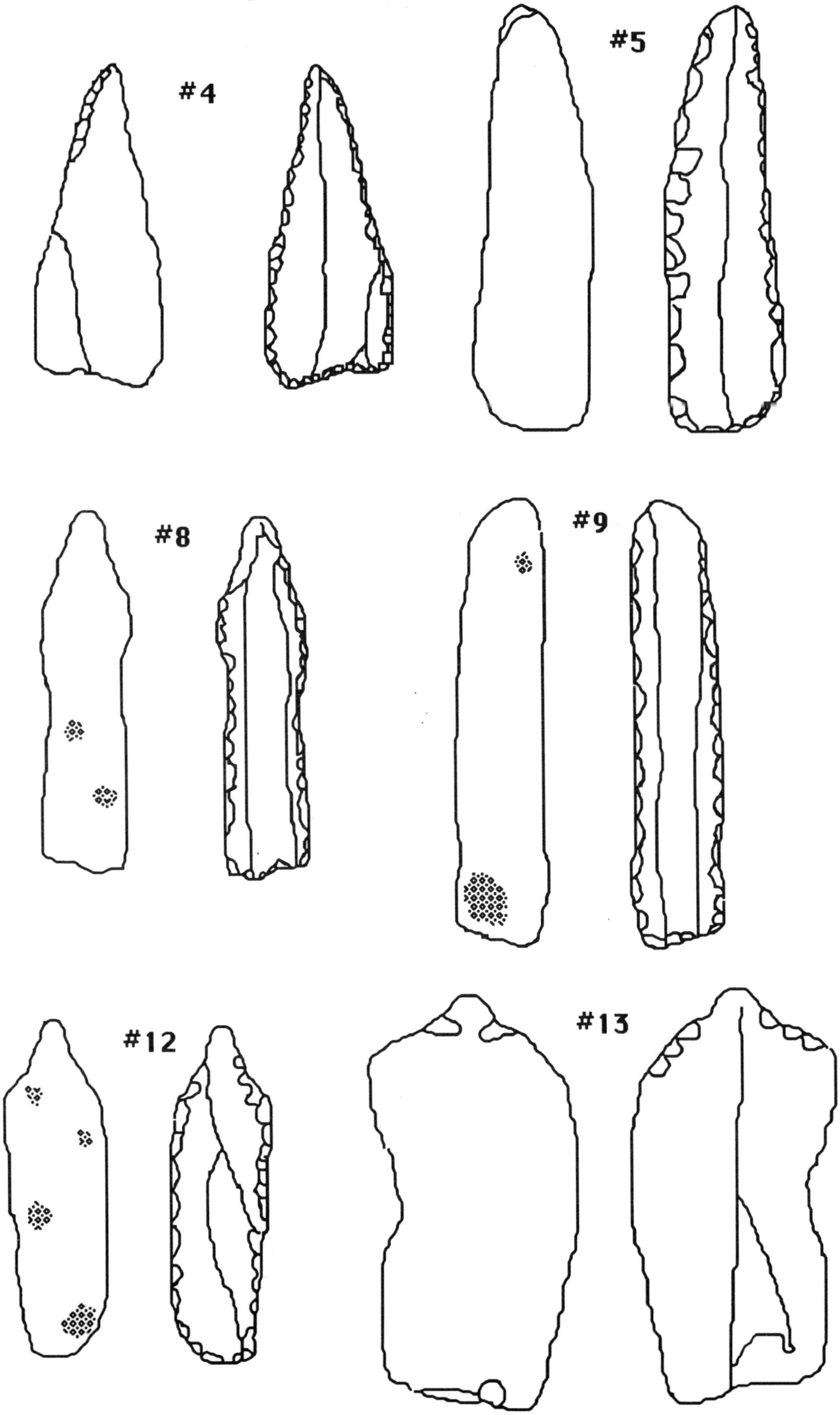

*

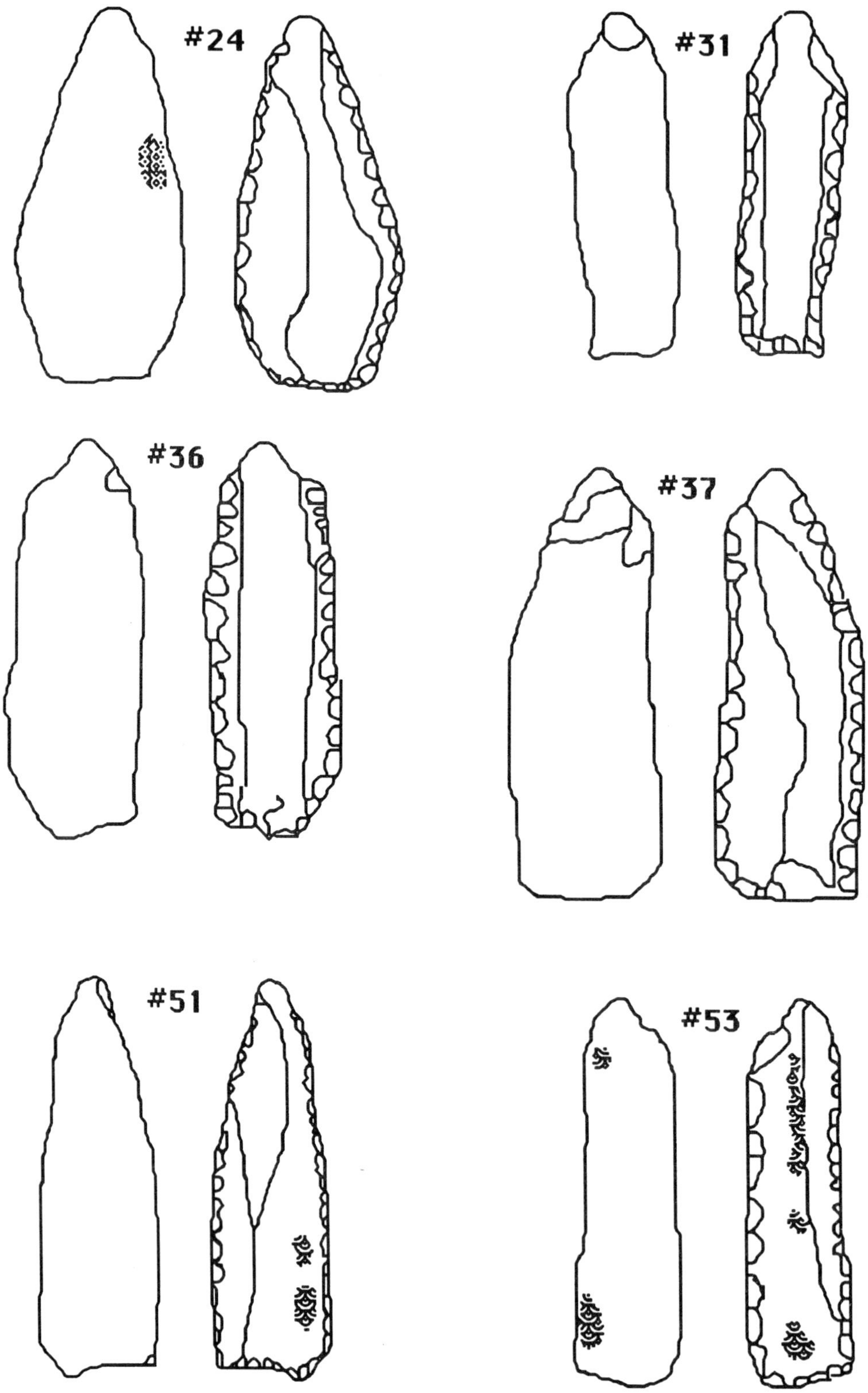

#24
#31
#36
#37
#51
#53

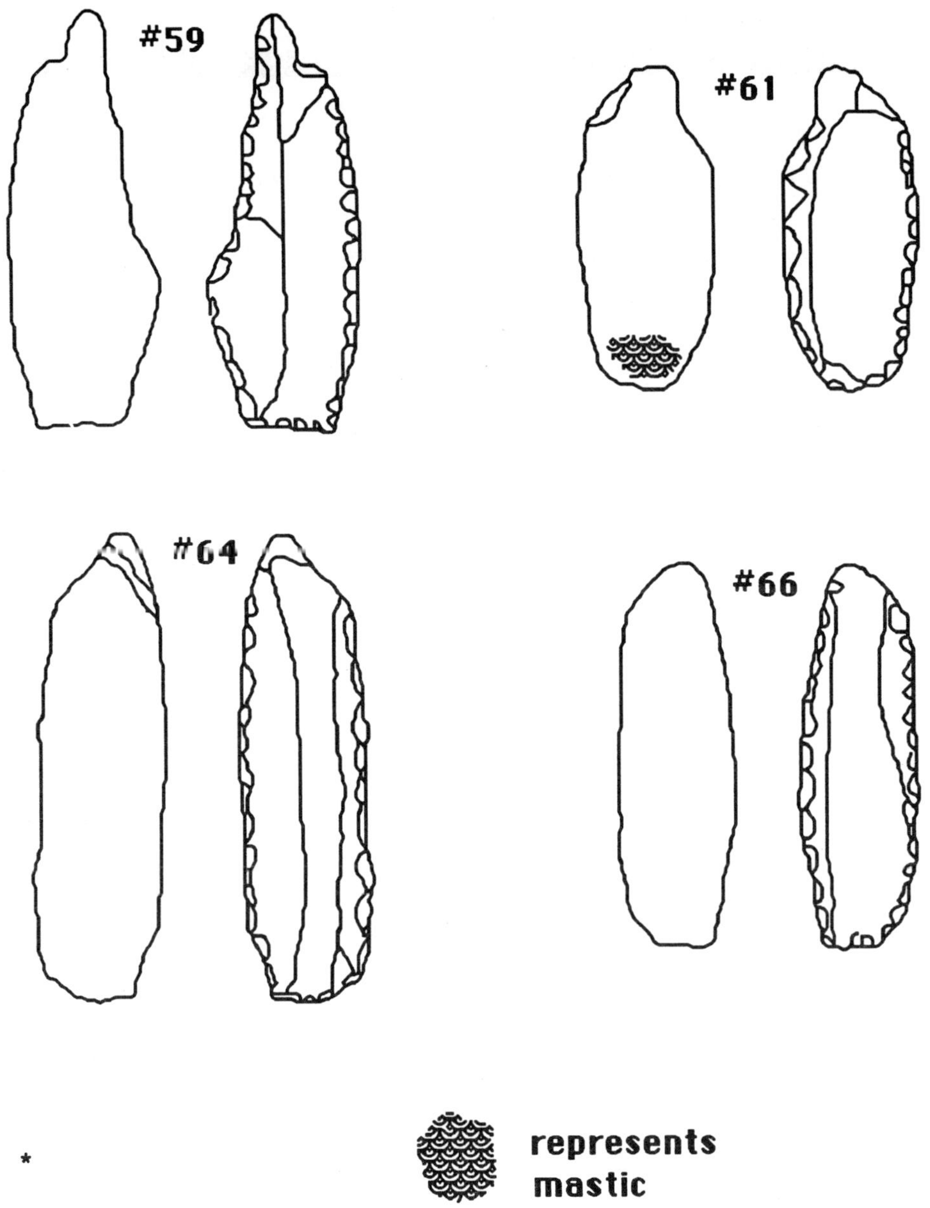
#59
#61
#64
#66
represents mastic

CHAPTER 6:
THE LIMITATIONS AND APPLICATIONS OF USE-WEAR ANALYSIS

One of the problems with microwear analysis is how to deal with post-depositional effects. This problem of differentiating between use-wear polish and post-depositional surface modification has been minimised by many microwear analysts. Most have claimed, following Keeley (1980) that it is relatively easy to distinguish between genuine use-wear polish and polish produced by a variety of post-depositional processes. Processes such as trampling, soil movement, rolling of the tools by water, the effect of sand blowing over the tools in desert environments etc. The impression given is that the effect of these processes on stone tools can easily be recognised and consequently dismissed. However the contention presented here that polishes are not distinctive to worked material implies that polishes produced by these post-depositional processes can look the same as polishes produced by use. If the visible differences between polishes are insufficient to distinguish between different worked materials, then the same problem occurs with distinguishing between polishes produced by contact with any worked material and polishes produced by post-depositional processes. This problem has been addressed by Levi Sala (1986). More recent research by Levi Sala has further suggested that polishes produced by replicating post-depositional processes are indistinguishable from use-wear polishes (Levi Sala pers. comm.),

The gross effects after deposition, such as when the whole tool is covered by a macroscopically visible sheen on both surfaces, may be distinguished from use-wear polish. Often this kind of surface alteration can effect most, if not all, of the tools in an assemblage, especially when the tools are found in sandy deposits which have been effected by wind action. Such as with the mesolithic assemblage from Hengisbury Head (Levi Sala 1986). In a case such as this, the assemblage is simply not susceptible to use-wear analysis that involves concentrating on evidence from looking at the polish. The real problem comes with tools that have only been affected marginally so that the use-wear polish is altered both in its appearance and distribution, or partially obscured by post-depositional polish.

The use of the multi-dimensional approach presented here is a partial solution to this problem, as any interpretations are made from a variety of attributes rather than concentrating on polish appearance. For example, if the use-wear polish has been affected by post-depositional processes so that its distribution is obscured, the requirement that the polish distribution correlates with other features such as edge morphology and edge wear would not be satisfied. The polish distribution would be *out of step* and this should inform the analyst that processes other than use have taken place. So that with the multi-dimensional approach, in order to interpret an edge as used, there has to be corroborative evidence of morphology and edge wear with the polish, all of which has to be consistent with a particular motion of use.

Some criteria may be used to differentiate use-wear polish from post-depositional polish. Use-wear will only be present on a particular edge and not distributed over different areas of the tool. When it is present on an edge a clear differentiation should be visible between where the polish ends and the unpolished surface begins. Often use-wear polishes diminish in development away from the used edge but the penetration of the worked material should limit this. If no clear boundary can be observed this indicates that the polish has been effected by post-depositional processes and in such a case it may be impossible to separate out the use-wear polish.

In many cases, however, post-depositional effects can be identified. Examples are the ability of the analysts to identify correctly the spots of polish and randomly orientated striations on test tool 47 as the product of trampling (see Chapter 4, Section 8.3). The drill bits from Kumartepe also had spots of polish and random striations which were clearly not associated with the use-wear polish on the tips of the tools. In this case the cause of the post-depositional features is not known. The fact that this material was screened means that post excavation processes may have been responsible. A more cautionary approach (as recommended by Levi Sala 1986) should be taken to the effect of post-depositional processes, and in many cases these effects will mean that high power microwear analysis, which concentrates on polishes, will not be applicable.

This is also sometimes the case when dealing with raw materials other than flint. Most of the work done on obsidian indicates that it does not become polished during use in the same way that flint becomes polished. The principle use-wear features being edge wear and striations, which are much more numerous on obsidian than on flint possibly because of the more brittle nature of obsidian. Polish seems to be for the most part absent on obsidian. The coarseness of materials such as quartzite and basalt often preclude the use of high power microscopy. The grain size precludes the observation of any polish that might be present because of the inability to focus on such material with optical microscopes at 200 or more magnifications. With such material the use of a SEM is probably more profitable (see Knutsson 1985).

In cases of gross post-depositional effects, or raw material that is not susceptible to high power microscopy, the use-wear analysis may have to be limited to edge attribute and edge wear analysis at low power. This kind of analysis is not able to provide sufficient information on which to base interpretations of precise worked material, but can identify used from unused tools and the hardness of the worked materials.

Post-depositional effects and raw material are two limitations that restrict the kind of use-wear analysis that may be undertaken on any particular assemblage. Another major restriction is the amount of time required for use-wear analysis and replication experiments that are necessary in order to produce reliable results. These limitations mean that the analysis of total assemblages with the intention of producing specific results, especially of worked materials, is not feasible. This means that to produce archaeologically significant results, use-wear analysis should be adapted in order to answer specific problems, rather than producing lists of activities of individual tools that rarely can represent the total assemblage from all but very small sites. The most profitable use of use-wear analysis is to apply it to specific problems within the general framework of a research strategy.

The approach to use-wear analysis presented in Chapter 4 enables the analysis to be used at different levels. There are three levels of analysis that can be carried out. The first is based on the morphological attributes of used edges and macro wear, which might be called *edge analysis*. The second level, in addition to edge analysis would include micro edge wear and rounding with the use of low power microscopy (*edge wear analysis*). The third level of analysis is to use both edge analysis and edge wear analysis in conjunction with high power microscopy looking at polish distribution (*microwear analysis*). The level of analysis that would be undertaken would depend on the condition of the material and the specific archaeological questions that are being asked of the material. Therefore the level of analysis would be determined by the information required in order to answer the specific problem.

For example, if one is concerned with variability between assemblages in terms of activities that took place at the sites, then edge analysis can group the tools into types having similar functional capabilities. That is, groups of tools are held to be associated with a particular activity, but without necessarily specifying exactly what that activity is. Therefore, one can construct a functional typology that can be quantitatively compared with other assemblages in order to ascertain the similarities or differences in activities represented at different sites. The problem of functional variability in the Mousterian assemblages of South West France is a case in point. Provided that the tool samples are appropriate, the contention that the variability is due to functional differences could be tested by discovering if the different assemblages had similar functional capabilities, without necessarily knowing the particular activity that is represented by a group of functionally similar tools. The question is, are these assemblages representing the same range of activities whatever those activities might be? As explained above the precise determination of the function of each tool in a number of large assemblages would not be possible. Therefore the use of a lower level of analysis, such as *edge analysis* which does not require microscopic examination, would be more appropriate to this situation. In this case quantity is more important than quality of information because the problem is concerned with statistical comparisons which require large samples. The tools may be grouped into functional types and their percentage representation compared with other assemblages in a similar way to which Bordean morphological types are used. These functional types can be derived by quantitative means such as cluster analysis (see Grace 1981). The difference between this and other attempts to cluster tools by function is that the unit of study here is that of used edges rather than the morphology of the whole tool and the placement of retouch. For example, the attempt by Binford and Binford

(1966) to produce tool kits by statistical association through the use of factor analysis was misguided, not only in the failure to interpret the statistics correctly but also because the basic data was of Bordean types which are not intended to be functional types : "the computer tells us that such and such tools covary; it does not say for what they were used' (Bordes 1972). Therefore, as the basic data are not primarily functional, the clustering of such types is not likely to produce functional associations. In Binford's factor analysis a side scraper is regarded as a functional type whereas the evidence from use-wear analysis strongly suggests that side scrapers were used for a number of activities and therefore cannot be considered as a functional type. (see Panagopoulou 1985).

Another way in which use-wear analysis can be used on whole assemblages is to attempt to interpret the function of a site as a whole, in terms of the range of activities that were carried out at the site. Knowing the range of activities would help to interpret the function of the site as a home base, kill site, hunting station, specialist activity site (such as a hide processing site) etc. For this kind of analysis the application of *edge wear analysis* could be used to ascertain the range of activities carried out at the site. *Edge wear analysis* would provide information about the motions of tools and the range of worked materials in terms of hardness. This information would be of the order of scraping soft and/or hard materials, boring/drilling, cutting soft and/or hard materials, use of projectile points etc. Therefore it could be established whether an assemblage represents a whole range of activities or concentration on particular activities. To establish the range and relative importance of activities it is not necessary to know the precise worked material on which each tool was used. Site interpretations from this kind of information could be that a whole range of activities would represent a base camp whereas a restricted range of activities would represent, for example, a hunting station or kill site if the activities were limited to the use of projectile points and cutting (butchering) tools. In such an analysis, having used *edge wear analysis* in order to separate the tools into groups associated with a particular kind of activity, samples from these groups could be taken and analysed using *microwear analysis* to obtain more precise information about the specific worked materials. When studying large assemblages a hierarchy of analysis would be necessary to overcome the problems mentioned above. *Edge analysis* would be preliminary to *edge wear analysis,* which would be preliminary to *microwear analysis* in order to sample the otherwise unmanageable numbers of tools when studying the whole assemblage. The sample size used for a full microwear analysis would be dependent on the results of the lower levels of analysis, so that the sampling strategy would be based on the relative importance of activities represented, rather than an arbitrary number of tools that could be analysed within the limitations of time and money available.

The kind of analytical process described above could be applied to an assemblage in order to interpret the subsistence strategy associated with a particular site. The separation of the tools into those used on soft or hard materials would give an estimation of the importance of vegetable resources as opposed to hunting resources. For example, if most of the tools were used for cutting soft materials and there was an absence of hunting tools such as projectile points and tools used for cutting hard materials such as bone, then the interpretation would be that the subsistence strategy was concentrated on the procurement of vegetable resources rather than hunting. Conversely, if there was an absence of tools used for processing soft materials, yet plentiful hunting equipment and butchering tools, then the emphasis would be on hunting as the main subsistence strategy associated with that particular site.

The kind of assemblage that would be suitable for all levels of use-wear analysis is when the assemblage is sufficiently small. This kind of assemblage could be from a closed context where a small group of tools may be associated with a burial, for example. Then microwear analysis, including the use of extensive replication experiments, is appropriate in order to obtain as precise information as possible. The analysis of a group of tools from the closed context of a burial could attempt to answer questions such as, do the tools represent the tool kit of the buried person, or are the tools ceremonial objects especially placed in the burial perhaps to denote the status of the individual? Another example of a small but significant assemblage would be where a concentration of tools is found on a site that appears to have some spatial significance, in that they may represent a specialised activity within a site, such as a hide processing area within a base camp.

A second area in which use-wear analysis can be profitably utilised is to approach specific problems associated with a particular tool type. For example, there has been extensive analysis on the use of projectile points (Barton and Bergman 1982, Newcomer and Moss 1982, Fischer et al. 1984) not only to establish whether typological projectile points were actually used as such, but on the kind of target that might have been involved. For example, are the different types of projectile point designed for use on specific targets? The contention that transverse arrow heads were designed for bringing down birds could be tested by use-wear analysis.

The whole typological group of microliths would provide an interesting area for use-wear analysis to discover if there were a functional association between microliths of the same type. For example, was a particular type of microlith consistently selected for use as barbs on projectiles as opposed to another type selected for hafting in composite knives used for cutting ?

Another example of where use-wear analysis could be applied to a particular tool type is that presented in Chapter 5 of the analysis of micro-borers from Kumartepe. This is an example of where use-wear analysis has been employed to determine the technology of carnelian bead production, which can be compared with the technology employed elsewhere for the same activity (Chevalier et al. 1982).

Also, use-wear analysis on particular tool types can help to resolve typological problems. For example, the stratigraphic sequence of Font-Robert points, truncated elements and Noailles burins found at Laugerie-Haute was interpreted by Peyrony as representing a chronological sequence and hence the nomenclature of Perigordian 5 a,b, and c. The subsequent discovery of all three types together in the same level at Le Flageolet demonstrated that this interpretation was incorrect. Therefore, the variation signified by the presence of these tool types in different levels at Laugerie-Haute was re-interpreted as implying a functional variation, in that the tools types were associated with a particular activity. The interpretation of these tools representing different activities is based solely on the morphology of the tools. This functional interpretation would be more credible if the precise functions of these tools were known. Also, knowing the precise activity that these tool types represent might help to explain their stratigraphic separation at Laugerie-Haute and their being contemporary at Le Flageolet.

Another application of use-wear analysis is to approach problems of technological association. For example, the use of Levallois technique was once regarded as culturally significant in that its presence denoted the Levallois Culture. The discovery of Levallois technique in contexts that are different typologically in other ways, and separated both chronologically and geographically, has led to the interpretation that Levallois is a technique rather than a cultural indicator. The use-wear analysis of Levallois flakes might explain this phenomenon if, for example, it was found that Levallois flakes, from whatever context, were always used for similar activities, or for different activities within a particular cultural or chronological context. It has been suggested that the Levallois technique was used in order to produce flakes with long cutting edges with the implication that this was for functional reasons. This could be tested by use-wear analysis. Another aspect of the use-wear analysis of Levallois debitage would be to ascertain whether Levallois points were projectile points or merely pointed Levallois flakes.

Provided that the material is suitable, use-wear analysis could be applied to the interpretation of Quina retouch, which is held to be a cultural indicator as the Levallois technique was once regarded; it being associated with Quina Mousterian *cultures*. However, Quina retouch is found in contexts separated geographically and chronologically in the same way as the Levallois technique. For example, Quina retouch is found, and in similar percentages, in the Yabrudian assemblagesof the Levant. Also, the step and scale flaking that signifies Quina retouch is found on tools from New Guinea and among Australian Aboriginal assemblages. So that the presence of Quina retouch by itself does not necessarily denote a *Quina culture* in the same way that the presence of Levallois technique is no longer regarded as denoting the presence of a *Levallois Culture*. The application of use-wear analysis to tools having Quina retouch from cross-cultural contexts, could establish whether Quina retouch is a technique to produce a particular tool associated with a specific activity. If this were the case then Quina technique would be regarded in the same way as Levallois technique, blade technique, pressure retouch etc. That is, as a way of knapping flint rather than a stylistic or ethnic association.

As mentioned at the beginning of this chapter use-wear analysis should be employed within a general archaeological strategy, not only in terms of attempting to answer specific questions with use-wear analysis but also incorporating use-wear analysis with other techniques. Use-wear analysis should not be seen as a technique that is intended to supplant existing methods of lithic analysis but to supplement lithic analysis as a whole. The application of use-wear analysis to the problem of Mousterian variability only becomes relevant because of the revelation of that problem from the typological analysis using Bordean methodology.

The examination of drill bits from Kumartepe is an example of use-wear analysis being used in conjunction with technological analysis (Calley and Grace 1988). The technological analysis of the material was carried out independently by Calley but led naturally to the question of how the drill bits were used. The manufacturing techniques used at Kumartepe involved a consistent strategy in order to produce a very specific tool type. The inference that this material represents a manufacturing site, derived from the use-wear analysis because of the advanced drilling techniques, is supported by the whole process of the manufacture of drill bits from raw material selection, through core reduction techniques to the use of an anvil. The inference is that these technological processes were designed to produce a specific tool for a specific task. The use of a burin-like technique in the core reduction sequence is an example of a specific process used in order to produce the desired blanks, a technique which is absent, for example, at Larsa which appears to be a small scale operation. The use of mechanical drilling for the production of carnelian beads is also absent at Larsa (Chevalier et al. 1982). Therefore at Kumartepe we have very consistent manufacturing process of the drill bits and of the beads, almost like a production line for a sophisticated industry probably intended for the manufacture of export goods.

Other techniques of lithic analysis can be used in conjunction with use-wear analysis. Re-fitting enables a group of tools that were made contemporaneously to form a sample for examination by use-wear analysis. Spatial analysis, again to provide an archaeologically significant sample for the use-wear analyst. It is also important to know the environmental background to the site from which a sample may come. Knowing the environmental resources available, such as the plant materials and potential prey, would help the microwear analyst to eliminate some worked materials. The example mentioned in Chapter 1 of the association of Quina scrapers with the working of hard wood was inferred from the presence of similar step and scale flaking on Australian Aboriginal hard wood working tools (Gould, Koster and Sontz 1971). This inference would not have been made if the environmental information (that hard wood was not available in the archaeological context) had been considered.

These are examples of other techniques that can help the use-wear analyst; the reciprocal is that the results of any use-wear analysis should be integrated with the results from these other techniques in order to interpret a site, or to research a particular problem, in as complete a way as possible. No one technique provides all the answers but each technique provides clues that help to reconstruct and understand the past. Use-wear analysis is a new and developing technique that can provide unique information about the past, as long as its limitations are appreciated and understood.

APPENDIX 1:TESTING FOR EFFICIENCY

Tools A1-5 were used for whittling a 3 cm. diameter hawthorn branch into a point. No attempt was made to control stroke length, contact angle or pressure exerted on the tool. Rather the tools were used in the most efficient way possible to achieve the desired result. All the tools were used for 10 minutes except for tool A3 with which the task was completed in only seven minutes. The tools are presented in order of efficiency as assessed by the resulting points on the wooden branch. Morphological variables of the edges of the tools; angle, length, thickness, profile and shape were measured (see Chapter 4, section 5 for a definition of the variables and a description of how they are measured).

From this kind of information the effect of variables on efficiency can be assessed. Some tendencies can be detected. An increase in efficiency is achieved with the more acute angled edges, and with the shorter edges. Thickness does not correlate with efficiency but as thickness refers to the support piece it would not be expected to be a sensitive diagnostic variable for tools used for whittling. The profile of the edges does correlate with efficiency, the convex edges being less efficient that the concave edges, particularly with the extreme concavity of the Clactonian notch (tool A3). From using the tools the increase in efficiency of concave tools for this specific task was due to the concavity *fitting* the piece of wood and preventing the force being applied from being dissipated by the edge sliding over the surface of the wood rather than cutting into it.

The shape variable does not appear to correlate with efficiency, though the longest/narrowest support piece (tool A5, shape = 2.88), may have been more efficient except for the denticulation of the edge which decreased the tools efficiency as it increased the edge angle (70°). The serrated edge was not suitable for a slicing action and tended to gouge the wood rather than cutting it cleanly. The next longest/narrowest tool (shape = 2.14) was A3 the Clactonian notch, whose handling properties (i.e. the relationship between support piece and the used edge) were excellent allowing a firm grip of the tool evenly around the working edge.

Though tool A3, by far the most efficient tool, had a relatively high edge angle (52°) the concavity of the notch (profile = -0.3) and the fact that the unretouched edge provided a clean cutting edge more than compensated for any deficiency due to the middle range edge angle. In fact all the unretouched edges (A4,A2,A3) proved more efficient than the retouched tools (A1,A5).

A 1 sidescraper

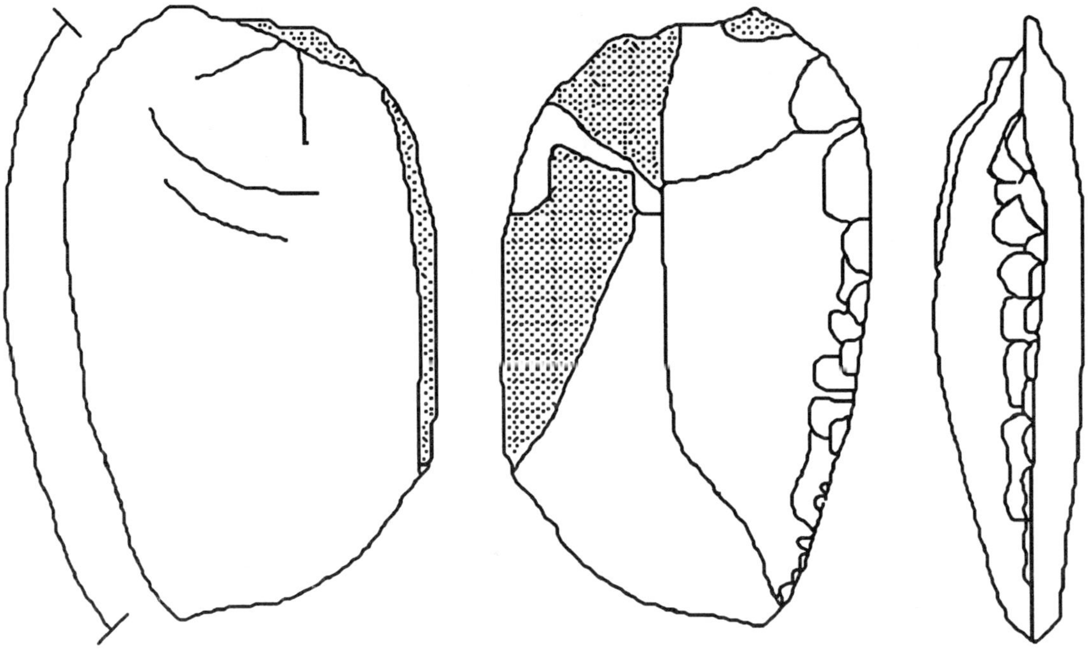

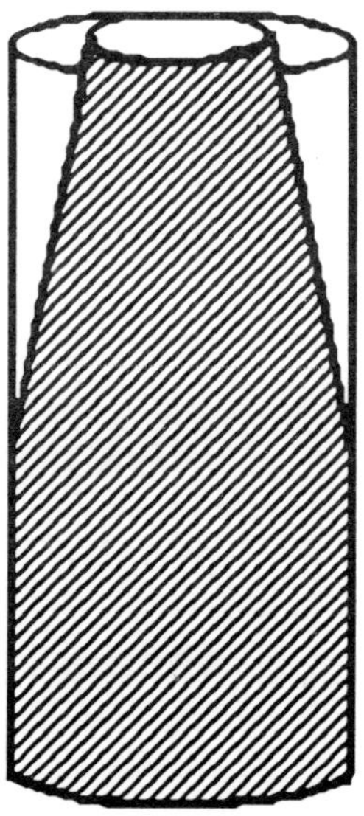

angle	60
length	101
thickness	23
profile	0.18
shape	1.82

A5 denticulate

*

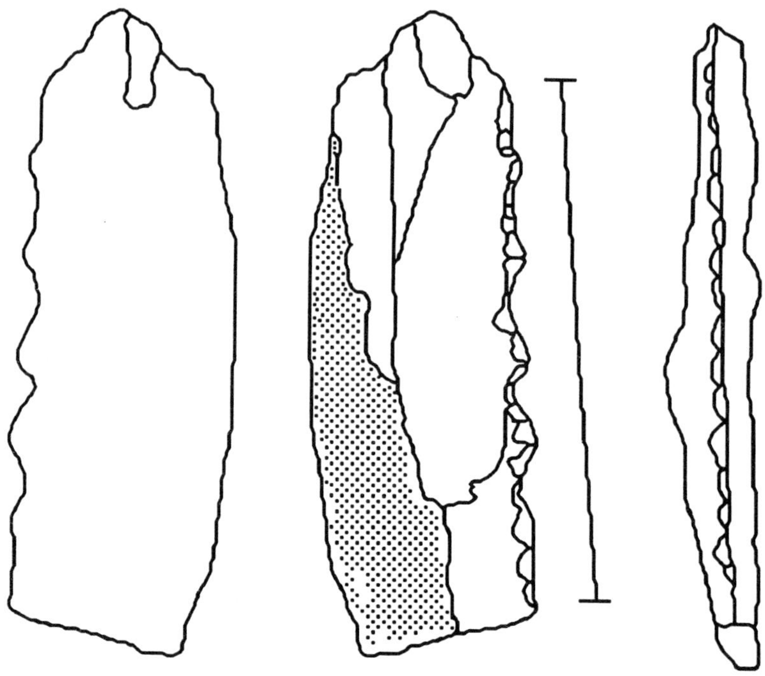

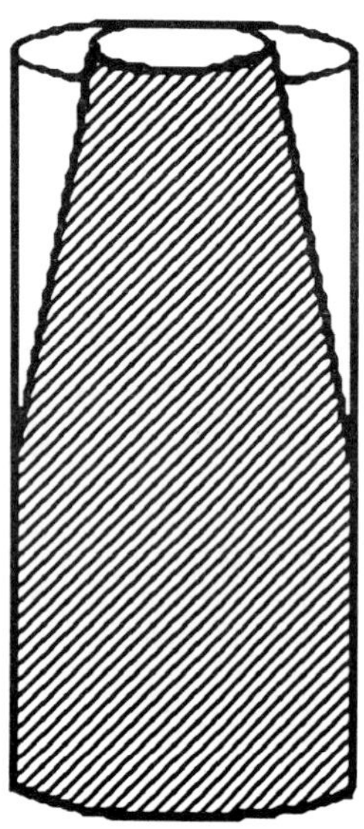

angle	40
length	54
thickness	13
profile	-0.11
shape	1.85

A 4 backed flake

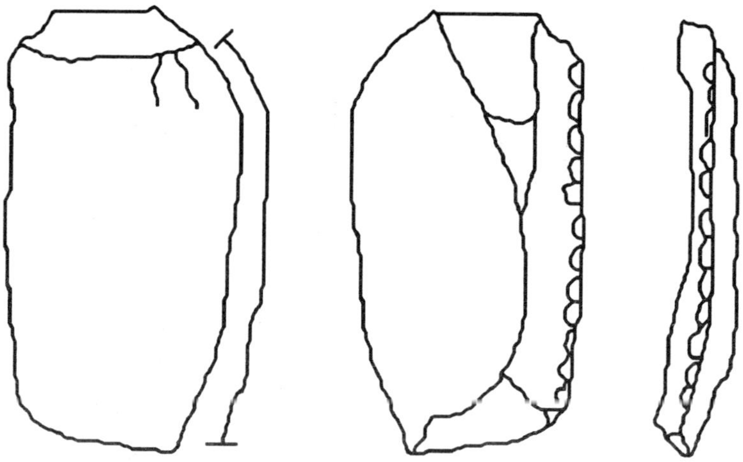

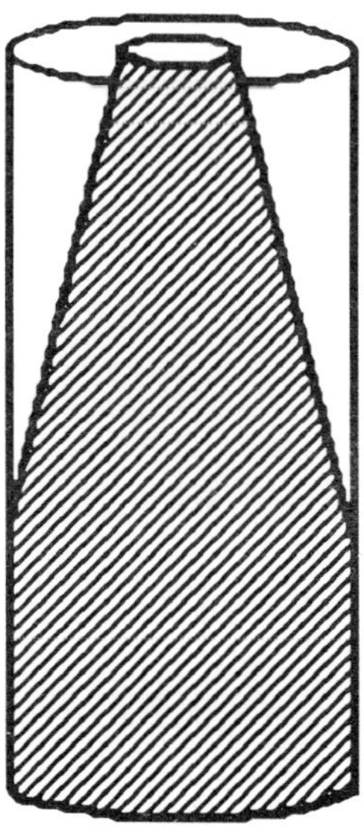

angle	38
length	60
thickness	14
profile	0.15
shape	1.29

A 2 backed flake

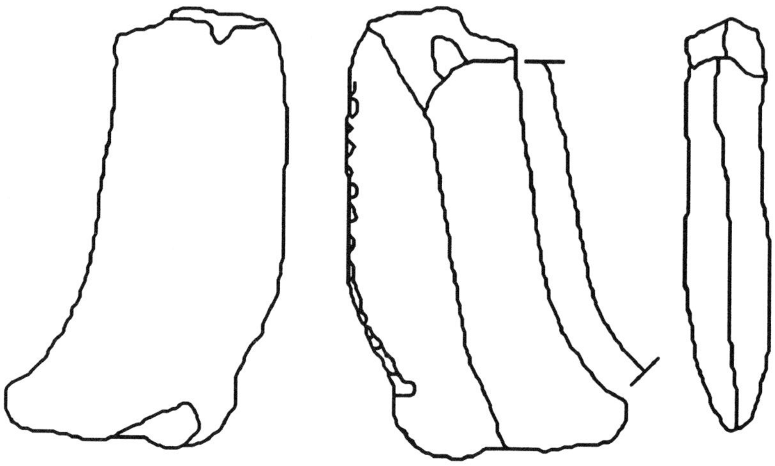

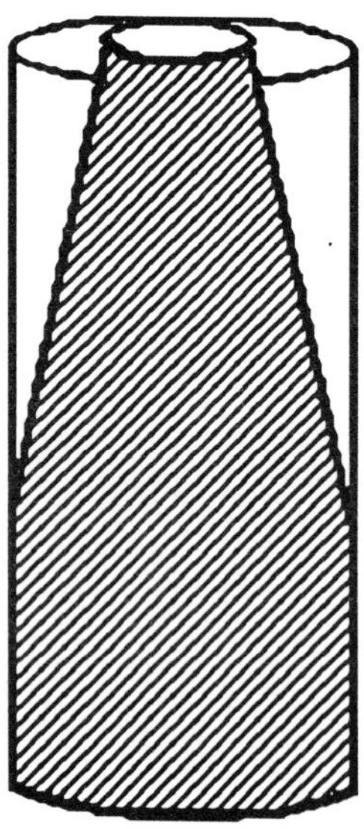

angle	70
length	81
thickness	10
profile	0.07
shape	2.88

A3 Clactonian notch

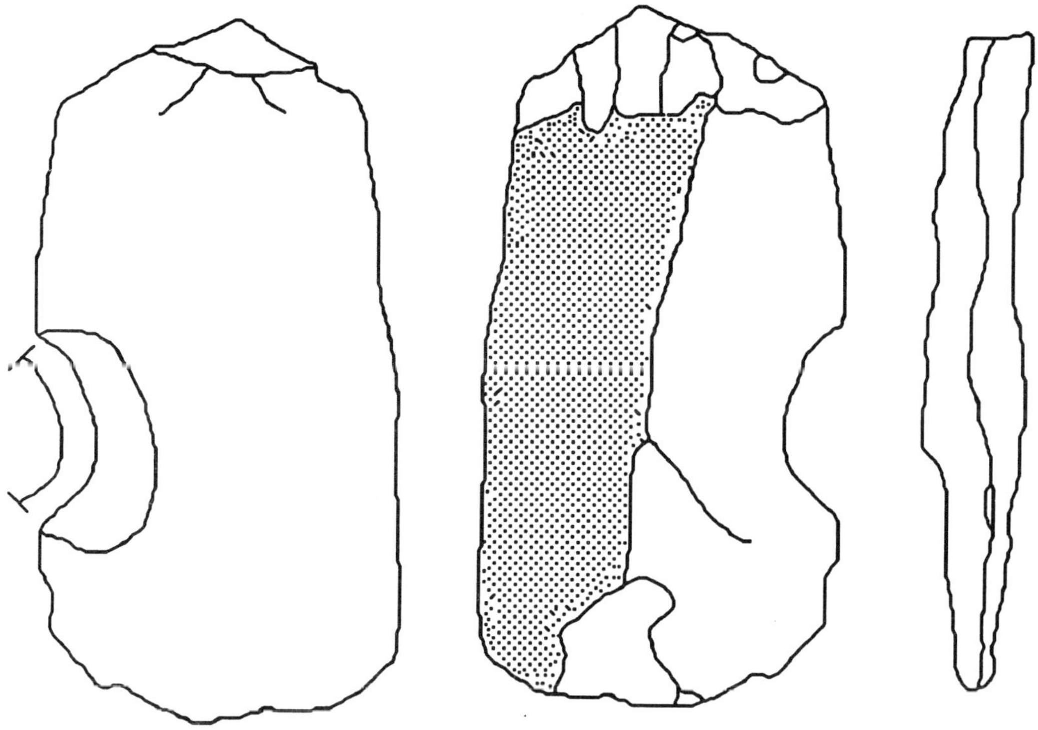

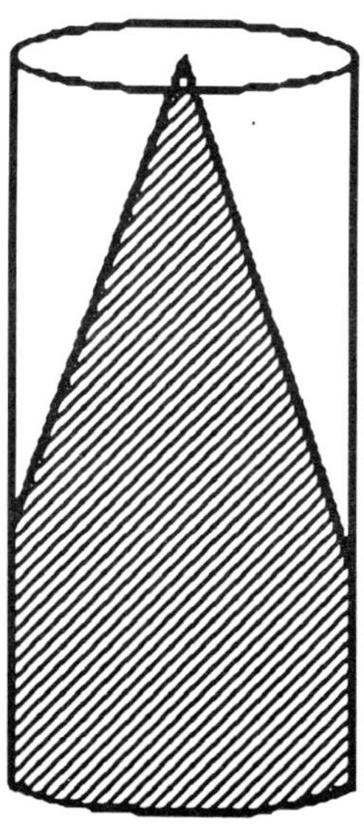

angle	52
length	38
thickness	22
profile	-0.3
shape	2.14

Tools A6-9 were used for sawing the same hawthorn branch for 10 minutes. The more acute angled edges tended to be more efficient. The relative inefficiency of tool A7, having an edge angle of 29°, was caused by the thickness of the support piece preventing the thin edge from penetrating into the wood. The most efficient tool (A9) had an acute angled edge (22°) and was also thin (thickness = 7). No concave profiled edges were used as this precludes their use as sawing tools as it is mechanically impossible to saw with a Clactonian notch, for example. There is a strong correlation between support piece shape and efficiency (i.e. the longer/narrower tools were more efficient). Tool A6 the transverse scraper, had invasive stepped retouch that not only increased the edge angle (82°) but also created a roughened surface that produced a wide, shallow cut. The unretouched edge of tool A7 became blunt quite quickly because the bending stress caused by the sawing motion created large snap fractures presenting a blunted edge to the wood. The denticulation of tool A8 proved more efficient but the increase in edge angle produced by the large scars detracted from its over all efficiency. Tool A9 though typologically a side scraper proved most efficient, having an acute angled edge (22°) and a thin, (thickness = 7) long and narrow support piece (shape = 2.18). The fine retouch on this tool did not significantly increase the edge angle but provided a more stable edge than that of the unretouched edge of tool A7.

This small scale experimental program indicates that there are correlations between the efficiency of tools, used for a specific task, with their edge attributes. The techniques presented here have been applied to a sample (n=252) of archaeological flints and the correlations between variables used as the basis for clustering the tools. The inference being that tools that cluster according to these variables are suitable (i.e. more efficient) for specific tasks (Grace 1981).

A6 transverse scraper

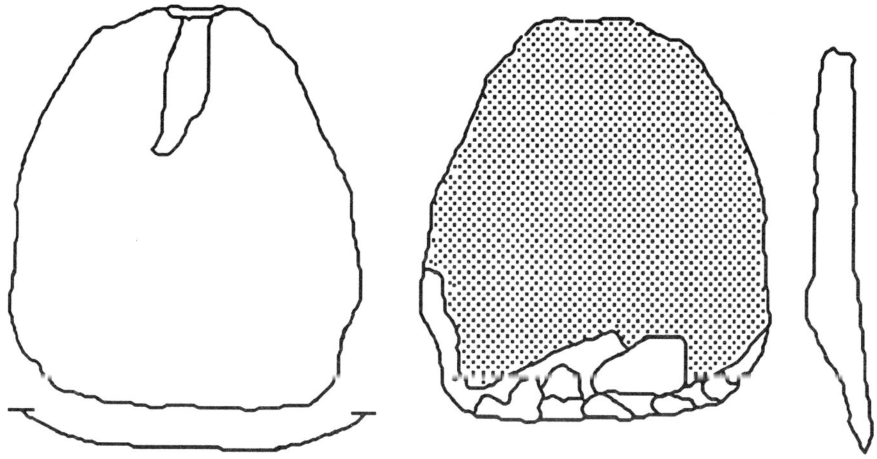

A7 flake

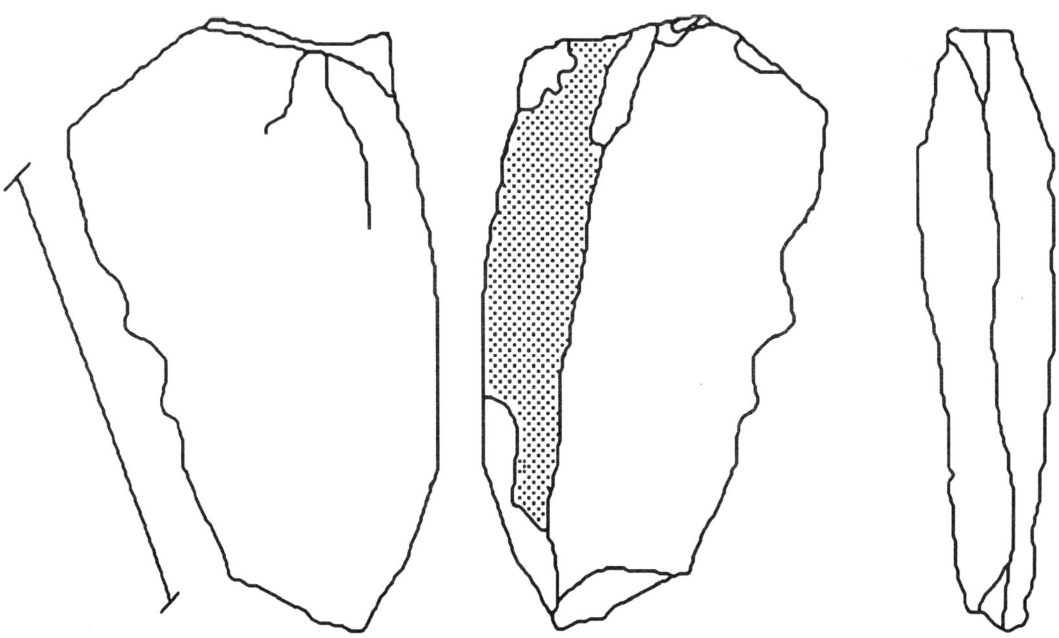

A8 denticulate

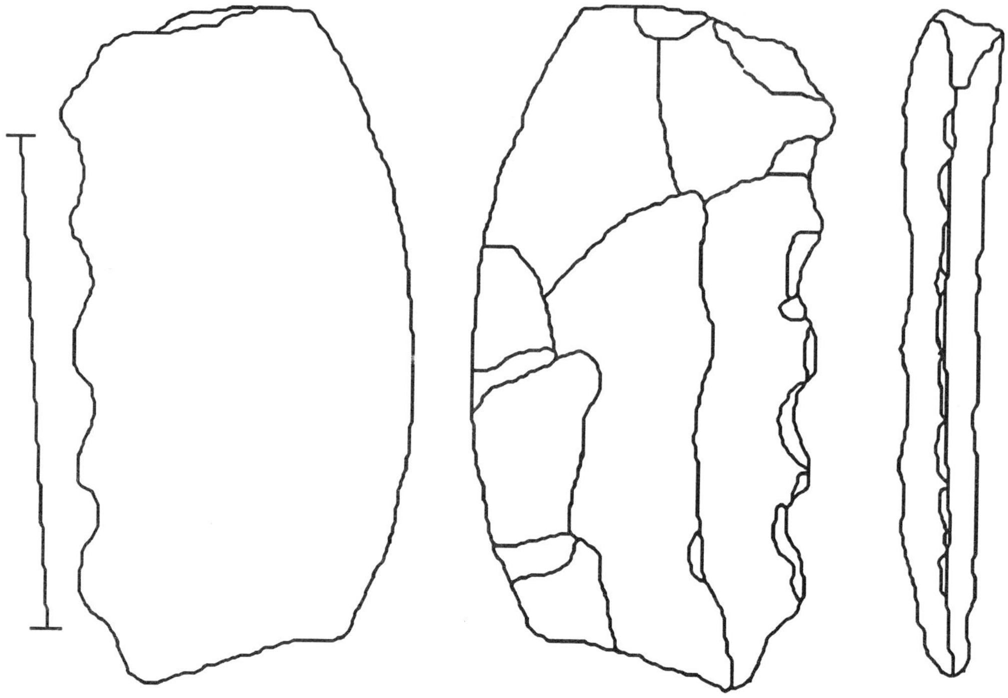

A9 sidescraper

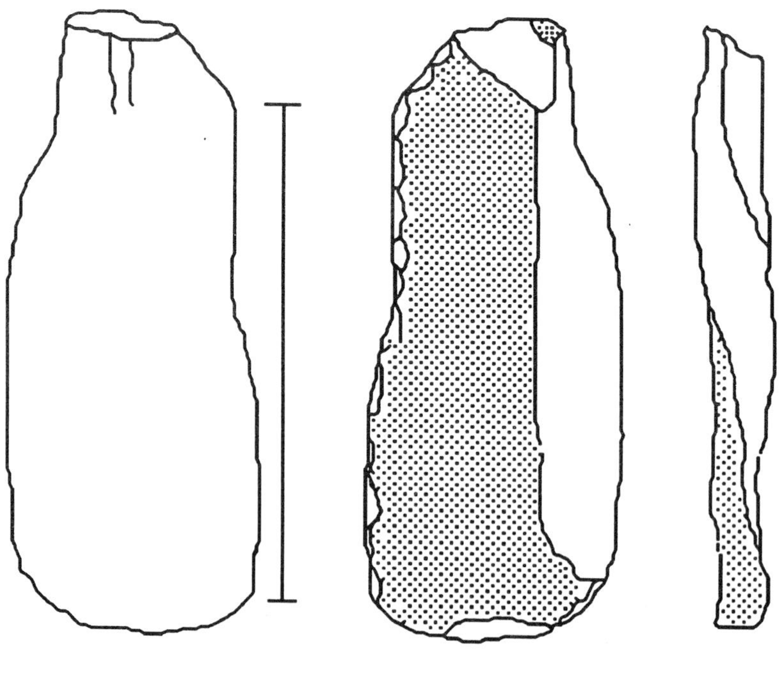

Amount of work achieved after 10 minutes sawing

	angle	length	thick	profile	
A6	82	56	10	0.16	0.93
A7	29	56	17	0	1.6
A8	57	61	10	0	2.06
A9	22	53	7	0.06	2.18

A6

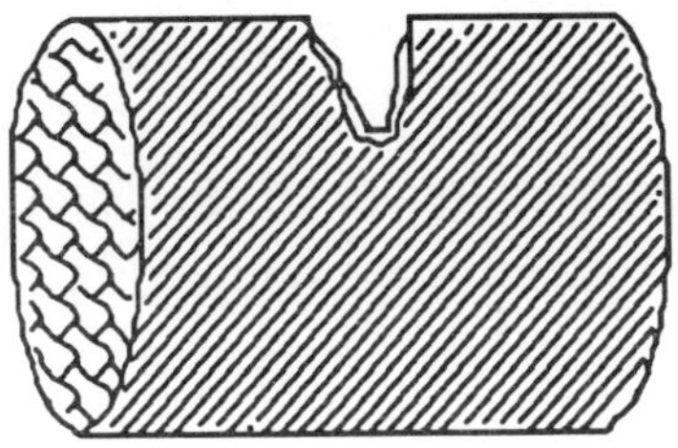

A7

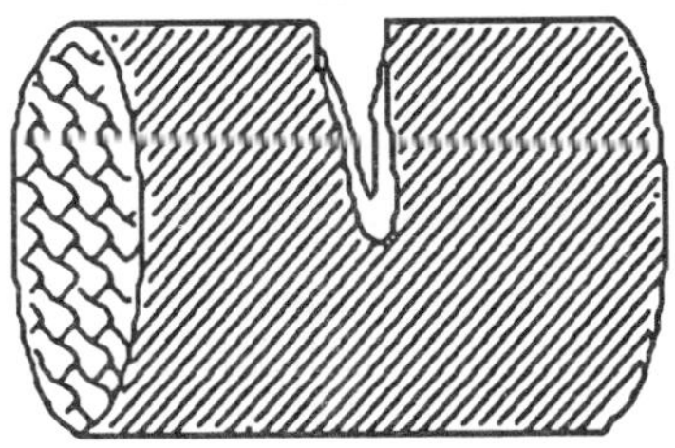

A8

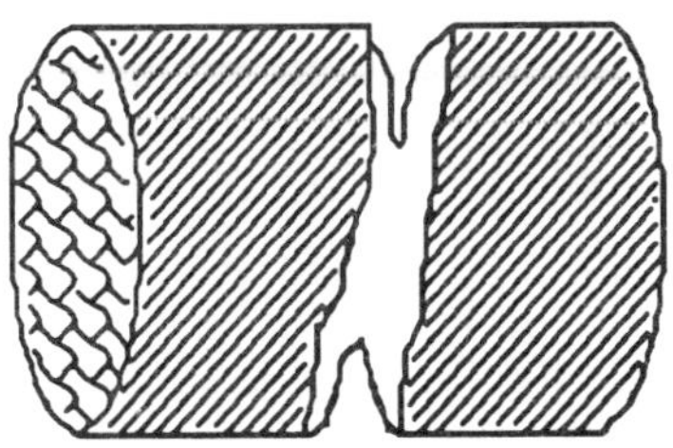

A9

In contrast to the experiments discussed in chapter 1, where the function of tools is assumed, these kind of experiments are designed to define the limits of a tool's capability. This in turn can indicate possible functions, or at least limit the range of functions that the tool is capable of.

APPENDIX 2:IMAGE PROCESSING STATISTICS

The experimental program on which the statistical results are based comprised of a group of tools mostly made and used by myself. To extend the experimental base I included tools made and used by other experimenters to ensure that any personal bias in techniques would not effect the data base. The areas were sampled by row, column co-ordinates to ensure only areas of polish were sampled, or for the unused flint examples, to ensure that no polish was present. Table 1 illustrates the values for the CON and ASM statistics expressed as differences from a *model* unused flint texture and a *model* polish produced by contact with antler. Tools prefixed **E** were made and used by myself. Tools prefixed **N** (and the blind test tools) were made and used by Mark Newcomer. The sickle blades prefixed **UH** were made by Newcomer and hafted in a sickle and used by Romana Unger-Hamilton. The drill bits were made by Christopher Bergman and used by him and myself.

The experimental program includes tools used in the following motions; sawing, whittling, scraping, grooving, drilling and a projectile. Worked materials include; wood, bone, antler, plants, hide, shell, stone, fish, meat and turquoise. The experimental data that this work is based on includes hundreds of tools made and used by myself or made available to me by my colleagues at the Institute, but only the ones actually digitally scanned for texture analysis have been reported here.

TOOL		TYPE	MOTION	MATERIAL		UNUSED CON	UNUSED ASM	ANTLER CON	ANTLER ASM
1	E3/F	flake	sawing	wood	15'	10.94	1.93	- - -	- -
2	E5/D1	flake	whittling	wood	20'	13.66	3.64	2.68	1.32
3	E5/D2	flake	whittling	wood	20'	14.81	3.03	- - -	- - -
4	E7/F	flake	sawing	bone	5'	16.46	4.25	3.93	1.89
5	E7/C1	flake	sawing	bone	5'	17.26	4.26	5.13	2.12
6	E7/C2	flake	sawing	bone	5'	14.77	2.62	2.56	0.62
7	E10/D	burin facet	scraping	bone	12'	14.48	2.86	1.21	0.47
8	E10/A	burin facet	scraping	bone	12'	9.09	0.86	5.01	2.18
9	E15/C2	burin facet	scraping	antler	4'	14.34	2.83	2.26	0.69
10	N47/1	blade	sawing	antler	3'	12.35	2.68	2.68	0.94
11	N47/2	blade	sawing	antler	3'	10.84	2.37	2.08	0.67
12	N47/3	blade	sawing	antler	3'	9.75	2.63	3.45	0.61
13	UH2/1	sickle blade	cutting	plants	5hrs	18.16	4.92	5.61	2.45
14	UH2/2	sickle blade	cutting	plants	5hrs	10.3	1.88	2.95	0.81
15	UH3/1	sickle blade	cutting	plants	5hrs	8.42	1.36	4.56	1.3
16	E2/D	flake	cutting	grass	20'	17.0	5.03	4.36	2.55
17	E13/A	endscraper	scraping	hide	28'	13.81	3.1	- - -	- - -
18	E37/D	flake	cutting	hide	10'	13.57	4.29	1.75	2.02
19	E23/C	bladelet	unused area			5.51	0.98	17.93	2.92
20	E8/A	flake	whittling	wood	10'	7.64	1.51	5.24	1.12
21	E5/D3	flake	whittling	wood	20'	12.98	2.31	0.86	0.34
22	E41/A	flake	whittling	wood	30'	11.93	2.54	1.07	0.45
23	E41/B	flake	unused area			2.49	0.49	14.48	2.96
24	E42/B	flake	unused area			3.55	0.51	15.76	2.99
25	E42/C	flake	sawing	wood	15'	18.8	4.81	6.2	2.33
26	E43/B	flake	unused area			3.56	0.52	15.99	2.52
27	E45/B	flake	unused			3.54	0.46	16.07	2.96
28	E46/A	tip of blade	grooving	antler	5'	15.2	3.39	2.57	0.94
29	E46/B	blade	unused area			1.49	0.45	11.59	2.22
30	E47/A	piercer	grooving	antler	5'	10.45	1.66	2/32	0.92
31	Flake 3		unused			2.14	0.26	11.5	2.43
32	Flake 3/A		unused			1.31	0.49	11.97	2.12
33	Flake 4		unused			3.28	0.56	15.08	3.05
34	Flake 6		unused			2.16	0.28	10.66	2.25
35	Flake 9/B		unused			3.14	0.51	- - -	- - -
36	Flake 10		unused			2.21	0.29	14.84	2.78
37	BT1/A	endscraper	scraping	wood	15'	6.9	1.55	5.87	1.32
38	BT1/E	endscraper	unused area			2.56	0.87	10.2	2.04
39	BT2/A	drill bit	dorsal ridge			10.26	1.55	2.67	1.09

						UNUSED		ANTLER	
	TOOL	TYPE	MOTION	MATERIAL		CON	ASM	CON	ASM
40	BT4/A	endscraper	unused area			1.43	0.79	11.39	2.28
41	BT4/B	endscraper	scraping	stone	6'	6.22	0.6	8.92	3
42	BT9/A	blade	unused area			2.35	1.17	14.84	2.56
43	BT9/B	blade	cutting	fish	20'	5.04	1.46	8.69	1.21
44	BT10/B	burin facet	scraping	wood	9'	5.94	1.16	6.74	1.31
45	BT5/A	burin facet	scraping	bone	11'	15.11	3.77	2.54	1.31
46	BT5/B	burin	unused area			2.79	1.54	14.68	2.4
47	BT3/A	piercer	grooving	antler	12'	7.34	1.2	6.22	1.89
48	BT3/C	piercer	unused area			2.01	0.51	11.83	2.09
49	BT2/C	drill bit	hafted area			15.36	4.47	2.74	1.99
50	BT6/C	Azilian point	shot into earth (tip)			11.24	3.57	2.02	1.26
51	BT6/D	Azilian point	unused area			1.53	1.2	13.48	2.38
52	BT7/A	endscraper	unused (retouched edge)			5.77	2.09	7.04	1.4
53	BT7/B	endscraper	unused area			7.28	1.42	5.57	1.26
54	BT8/B	bladelet	cutting	meat	29'	2.32	0.83	14.8	2.25
55	BT8/C	bladelet	unused area			2.13	0.28	14.72	2.15

HAFTING EXPERIMENTS ON DRILL BITS

				UNUSED	
		BLANK/RETOUCH/DRILL	SCANNED AREA	CON	ASM
56	HE1/B	blank	haft	7.52	1.36
57	HE1/E	retouched	haft	1.13	0.31
58	HE1/F	retouched	tip	3.22	0.67
59	HE1/J	drill bit	tip	2.59	0.43
60	HE2/A	blank	haft	0.77	0.75
61	HE2/G	drill bit	haft	4.02	0.41
62	HE2/J	drill bit	dorsal ridge	8.93	2.15
63	HE3/B	blank	haft	0.77	0.55
64	HE3/I	drill bit	haft	1.33	0.44
65	HE3/M	drill bit	used tip	18.31	6.27
66	HE3/P	drill bit	used tip	24.87	24.25
67	HE3/Q	drill bit	used tip	10.79	1.86
68	HE4/A	blank	tip	2.16	0.69

		BLANK/RETOUCH/DRILL	SCANNED AREA	UNUSED CON	ASM
69	HE5/A	blank	érailler scar	4.19	1.04
70	HE5/D	blank	haft	1.13	0.85
71	HE5/K	drill bit	haft	1.29	1.06
72	HE5/N	drill bit	used tip	16.23	3.91
73	HE5/M	drill bit	broken tip	22.9	10.09
74	HE5/O	drill bit	broken tip	24.5	12.07
75	HE6/A	blank	haft	1.13	0.31
76	HE6/F	retouched	haft	1.77	0.26
77	HE7/D	blank	haft	0.55	0.87
78	HE7/E	retouched	haft	2.98	0.34
79	HE7/G	retouched	haft	24.78	26.57
80	HE8/A	blank	haft	1.32	0.22
81	HE8/F	retouched	haft	3.34	0.6
82	HE8/I	retouched	haft	6.74	1.26
83	HE8/J	retouched	haft	22.4	12.45
84	HE8/L	retouched	haft	1.99	0.97
85	HE9/E	retouched	haft	3.4	0.67
86	HE9/H	drill bit	haft	5.21	1.21

To illustrate the positions of the samples on the tools a selection of drawings are presented, including all the blind test tools and the drill bits used in the hafting experiments. The way in which the tools were used and on which material is shown in the above table.

The used edges of the tools are indicated by a line.

Cortex is indicated by stippled areas.

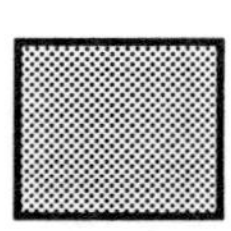

The scale of all the drawings is 1:1.

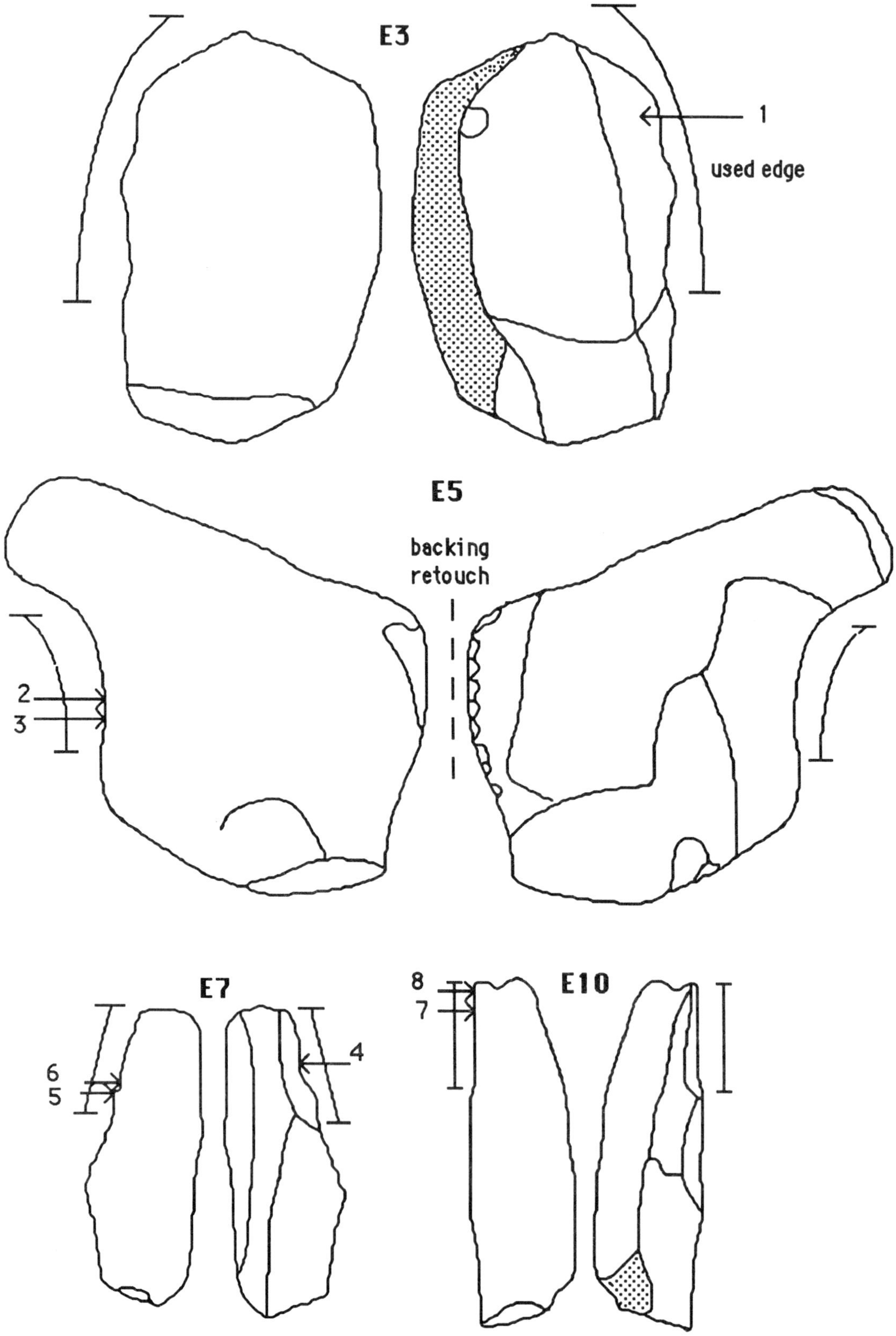
E3
1
used edge
E5
backing
retouch
2
3
E7
6
5
4
8
7
E10

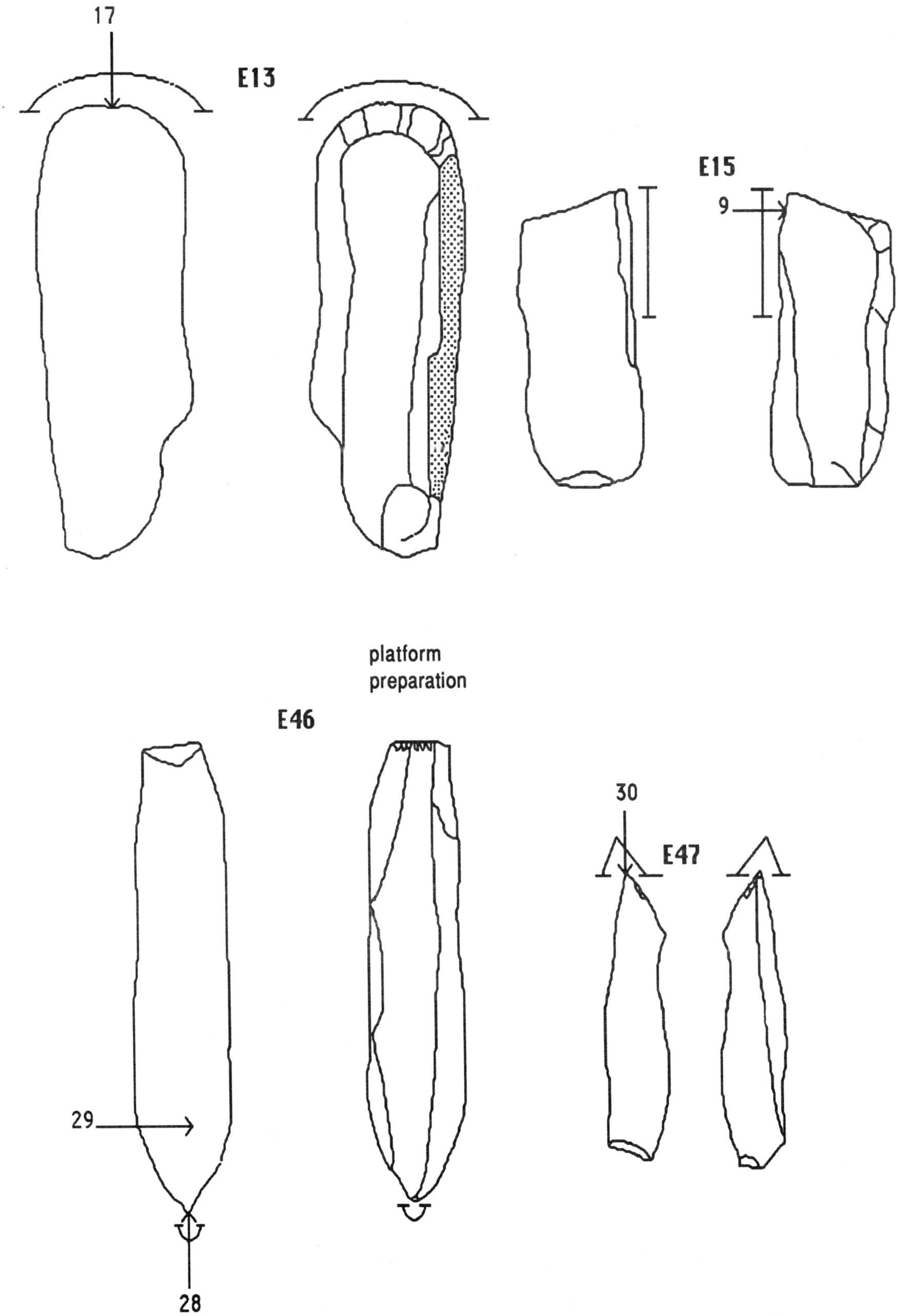
17
E13
E15
9
platform
preparation
E46
30
E47
29
28

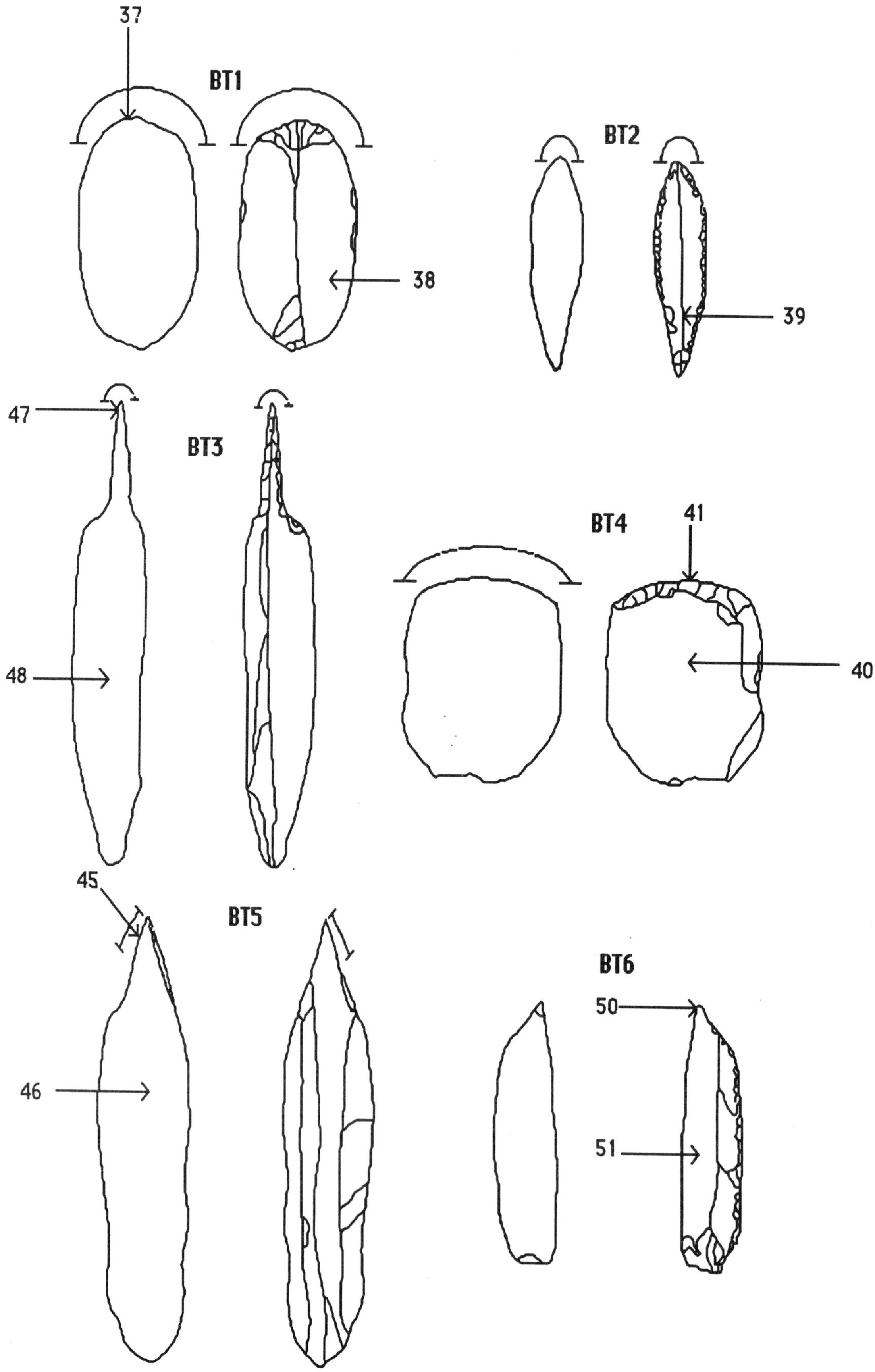
BT1
37
38
BT2
39
BT3
47
48
BT4
41
40
BT5
45
46
BT6
50
51

BT7

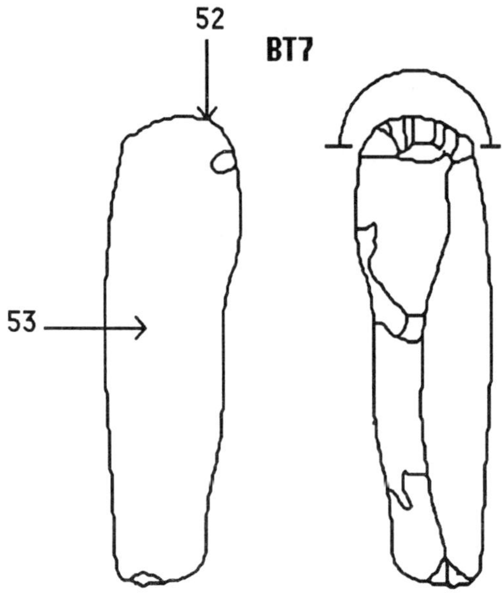

BT8

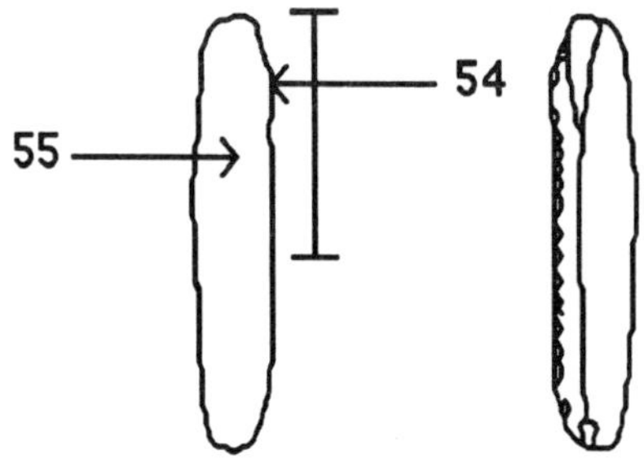

BT9

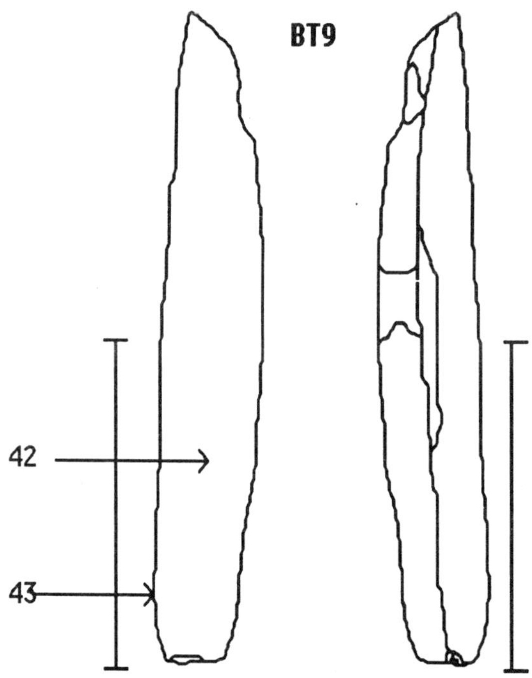

BT10

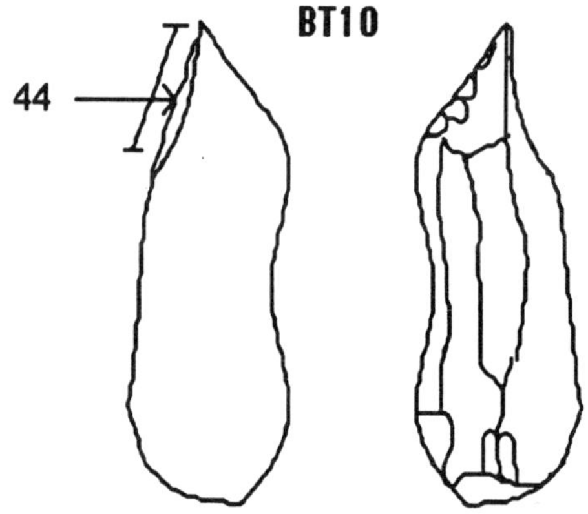

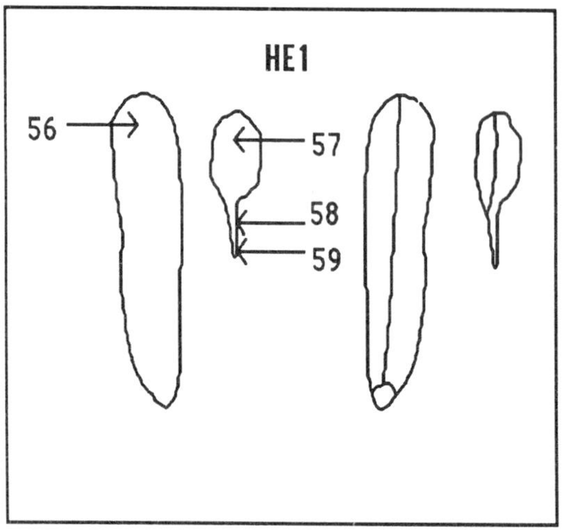
HE1
56
57
58
59

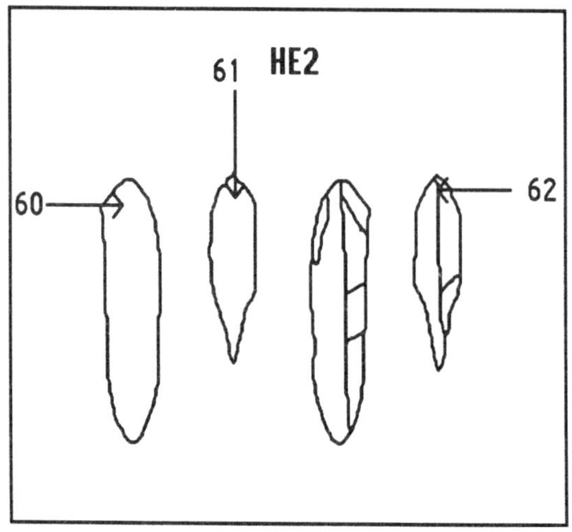
HE2
61
60
62

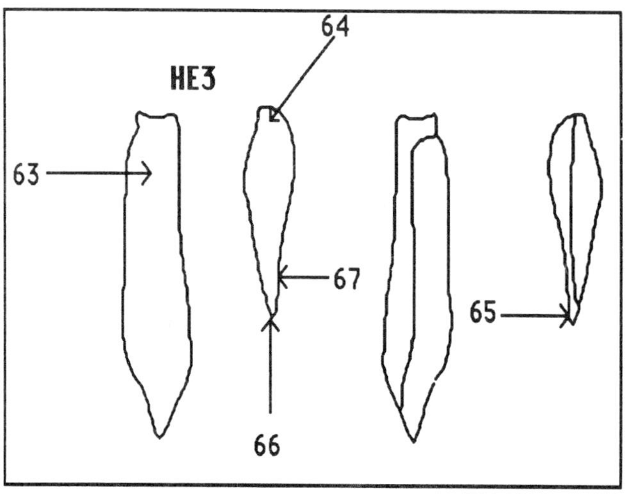
HE3
64
63
67
65
66

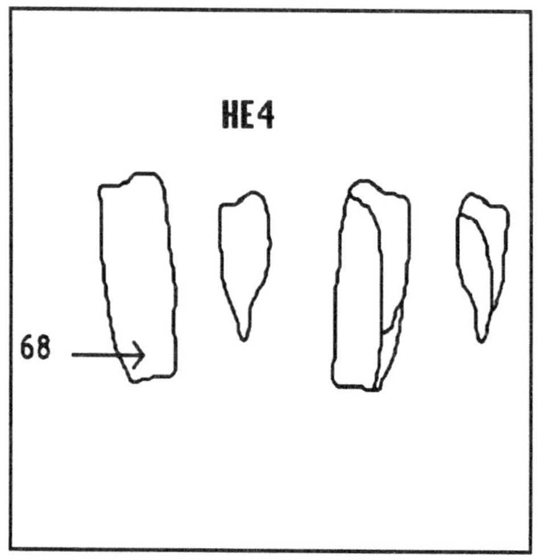
HE4
68

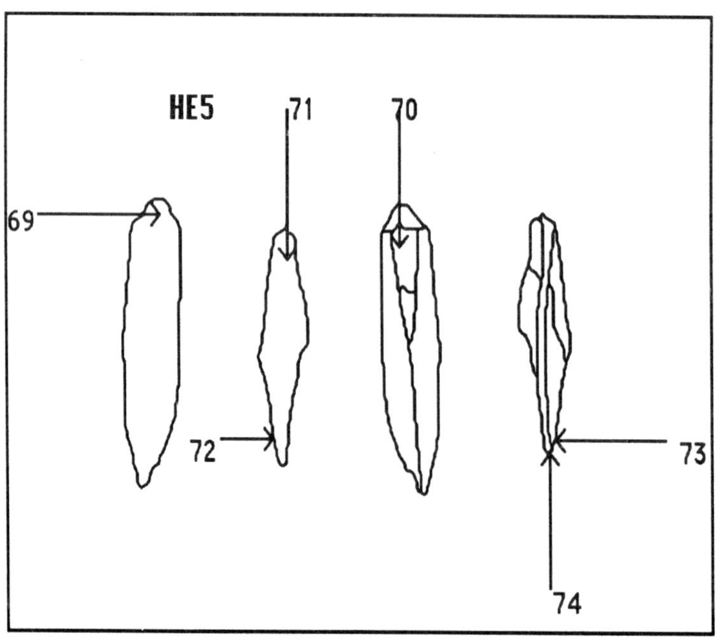
HE5
71
70
69
72
73
74

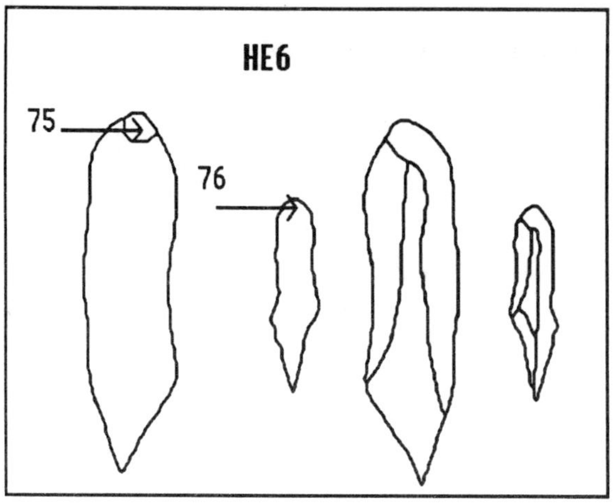
HE6
75
76

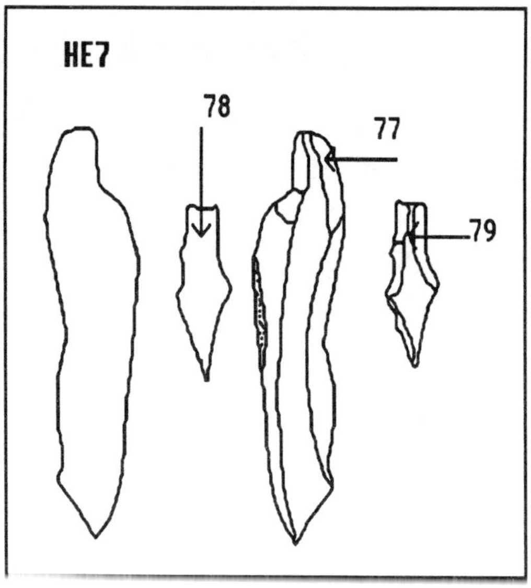
HE7
78
77
79

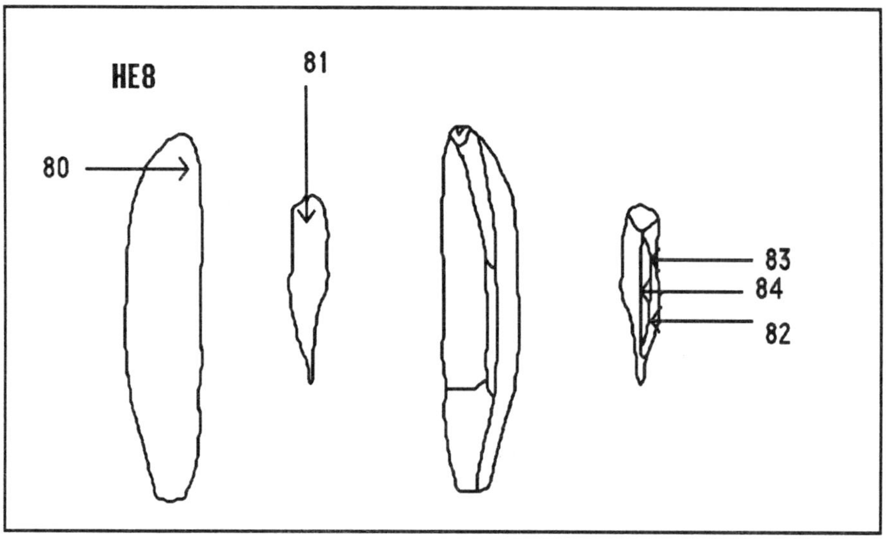
HE8
81
80
83
84
82

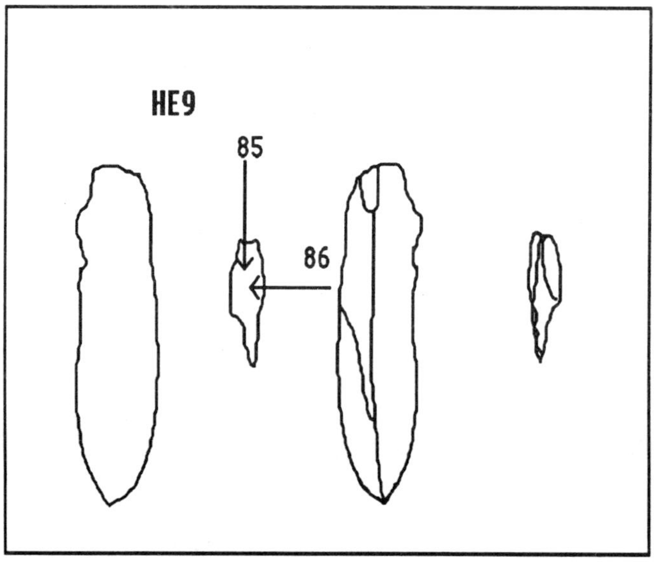
HE9
85
86

APPENDIX 3:BLIND TEST DOCUMENTATION

This appendix presents the full documentation for the blind test with my observations and interpretations. Outline drawings of each tool are illustrated with the edge that was interpreted as used marked on. These interpreted used edges are all correct except for tool 38 which was interpreted as unused whereas the burin tip was used to push holes in bark.

The following is a summary of the variables recorded on the observation sheet. The variables are presented in CAPITALS and the possible values in **bold**. On the sheets presented here 0 is used to denote absence.

GRAIN SIZE: A visual inspection of the tool to estimate whether the flint is **fine**, **medium** or **coarse** grained. [see Chapter 4, Section 5.1.1]

TOPOGRAPHY:A visual inspection of the ventral (or unretouched) surface adjacent to the used edge. The topography can be **flat, undulating** or **ridged.** [see Chapter 4, Section 5.1.2]

TOPOGRAPHIC FEATURES: A recording of the presence of **percussion ripples** or **edge feathering.** [see Chapter 4, Section 5.1.3]

EDGE ANGLE: The angle of the used edge measured in **degrees**. [see Chapter 4, Section 5.2.1]

LENGTH: length of used edge in **mm**. [see Chapter 4, Section 5.2.2]

THICKNESS: Thickness of piece at mid point of used edge in **mm**. [see Chapter 4, Section 5.2.3]

PROFILE: The curvature of the edge. O= straight, negative scores = concave edges. [see Chapter 4, Section 5.2.4]

SHAPE: The length/breadth ratio of the support piece with the tool in it's working orientation. [see Chapter 4, Section 5.2.5]

MACRO EDGE WEAR:

GLOSS: **Present** or **absent.** [see Chapter 4, Section 5.3.4]

FRACTURES: < or > 5 per 10mm. of used edge. [see Chapter 4, Section 5.3.1]

FRACTURE TYPE: **Flakes, snaps** or **steps.** [see Chapter 4, Section 5.3.2]

ROUNDING: **Light** or **heavy.** [see Chapter 4, Section 5.3.3]

MICRO WEAR: Observed at 200 magnifications.

EDGE WEAR as for macro except number of flakes is taken as **<** or **>** 5 per 5mm. [see Chapter 4, Section 5.4]

MICROTOPOGRAPHY: Topography of surface adjacent to used edge can be **flat, undulating** or **ridged.** [see Chapter 4, Section 5.5]

POLISH DISTRIBUTION: **Continuous** or **intermittent.** [see Chapter 4, Section 5.6]

POLISH DISTRIBUTION TYPE: Can be **away from the edge, gapped, edge only/even, edge only asymmetric** or **differential** [see Chapter 4, Section 5.7]

INVASIVENESS: The distance that the polish extends away from the edge. Can be **edge only** or **<0.5D i.e. <400μm** or **> 0.5D i/e. >400μm.** [see Chapter 4, Section 5.8]

LINEAR FEATURES: The orientation to the used edge of lines of polish. They can be **parallel, perpendicular, angled** or in any combination. [see Chapter 4, Section 5.9]

STRIATIONS: The orientation to the used edge of striations. They can be **parallel, perpendicular, angled** or in any combination. [see Chapter 4, Section 5.10]

POLISH DEVELOPMENT: The level of development of the polish from **A to C,** or the special case of linear polish **D.** [see Chapter 4, Section 5.11]

BLIND TEST TOOL #31

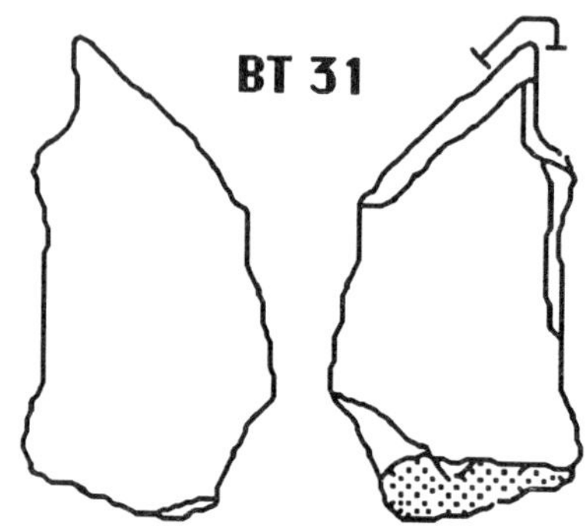

TYPE:	BURIN
USED AREA:	BURIN BIT
	morphology of tool and presence of edge wear and polish
DIRECTION OF USE:	TRANSVERSE
	across the bit due to the presence of edge wear
MOTION OF USE:	GROOVING
	logical motion of a short edge such as a burin bit, taking into account the lack of penetration of the material (invasiveness = edge only)
HARDNESS OF MATERIAL:	MEDIUM
	if a hard material more macro edge wear (=<5) would be expected also there was only light rounding
NAMED MATERIAL:	WOOD/SOAKED ANTLER
	harder of the materials more likely because of fractures, with wood would expect little or no edge wear
PROBABLE FUNCTION:	GROOVING ANTLER
ACTUAL USE:	GROOVING HORN for 5 minutes

For discussion of horn as a worked material see Chapter 4, Section 10.

OBSERVATIONS

TOOL NO.	31	
TYPE	dihedral burin bit	

EDGE MORPHOLOGY

EDGE ANGLE	46
EDGE LENGTH	6
THICKNESS	4
PROFILE	0.76
SHAPE	2.08

MACRO EDGE WEAR	VENTRAL	DORSAL
FRACTURES	<5	absent
FRACTURE TYPE	flakes	absent
ROUNDING	absent	absent
GLOSS	absent	absent

MICRO EDGE WEAR		
FRACTURES	>5	absent
FRACTURE TYPE	flakes	absent
ROUNDING	absent	absent
MICROTOPOGRAPHY OF POLISHED AREA	flat	flat
MICRO-POLISH DISTRIBUTION	intermittent	intermittent
DISTRIBUTION TYPE	edge only even	edge only even
INVASIVENESS	edge only	edge only
LINEAR FEATURES	absent	absent
STRIATIONS	absent	absent
POLISH DEVELOPMENT	B	B

BLIND TEST TOOL #32

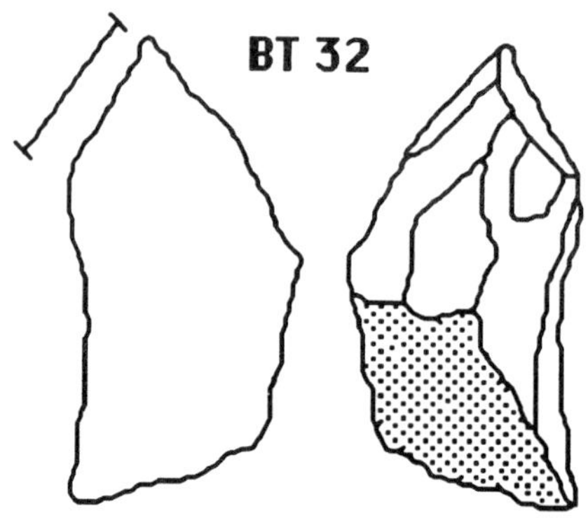

TYPE:	DIHEDRAL BURIN
USED AREA:	BURIN FACET
DIRECTION OF USE:	TRANSVERSE
	unsuitability of edge for longitudinal motion (edge angle=69)
MOTION OF USE:	SCRAPING
	edge wear indicated an across the edge motion to create flake fractures
HARDNESS OF MATERIAL:	SOFT
	well developed polish (type C) indicates a considerable period of use, if a harder material more edge wear would be expected
NAMED MATERIAL:	HIDE
	from above
PROBABLE FUNCTION:	SCRAPING HIDE

ACTUAL USE: SCRAPING ANTLER for 5 minutes

The well developed polish was interpreted as being due to the tool being used for a long period on a soft material whereas this was due to to a short period of use on a hard material. On reflection the amount of micro edge wear (>5 on ventral aspect of facet) should have been interpreted as caused by use on a harder material. This was an isolated mistake as the the other analysts involved in the test did not make this mistake, two of them identifying the worked material as antler and one as bone.

OBSERVATIONS

TOOL NO.	32	
TYPE	dihedral burin facet	
EDGE MORPHOLOGY		
EDGE ANGLE	69	
EDGE LENGTH	16	
THICKNESS	7	
PROFILE	0.1	
SHAPE	1.3	
MACRO EDGE WEAR	VENTRAL	DORSAL
FRACTURES	absent	<5
FRACTURE TYPE	absent	flakes
ROUNDING	absent	absent
GLOSS	absent	absent
MICRO EDGE WEAR		
FRACTURES	>5	<5
FRACTURE TYPE	flakes	flakes
ROUNDING	light	absent
MICROTOPOGRAPHY OF POLISHED AREA	flat	flat
MICRO-POLISH DISTRIBUTION	continuous	continuous
DISTRIBUTION TYPE	edge only even	edge only even
INVASIVENESS	<0.5	edge only
LINEAR FEATURES	absent	absent
STRIATIONS	absent	absent
POLISH DEVELOPMENT	C	C

BLIND TEST TOOL # 33

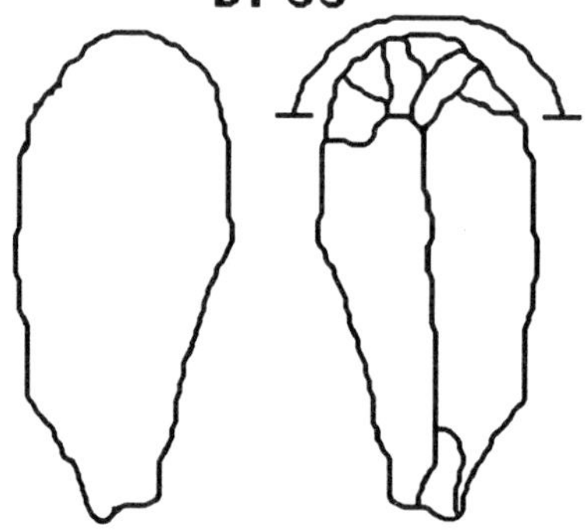

TYPE:	ENDSCRAPER
USED AREA:	RETOUCHED EDGE
	morphology, micro edge wear and polish
DIRECTION OF USE:	TRANSVERSE
	perpendicular linear features and striations
MOTION OF USE:	SCRAPING
	morphology, linear features and striations
HARDNESS OF MATERIAL:	HARD
	lack of penetration of material (invasiveness = >0.5D) and presence of linear features and striations
NAMED MATERIAL:	BONE/STONE
	deep striations indicate harder of materials
PROBABLE FUNCTION:	SCRAPING STONE

ACTUAL USE: SCRAPING HIDE for 15 minutes

Linear features and striations due to manufacture hence misinterpretation

OBSERVATIONS

TOOL NO.	33
TYPE	endscraper

EDGE MORPHOLOGY

EDGE ANGLE	42
EDGE LENGTH	28
THICKNESS	6
PROFILE	0.36
SHAPE	0.53

MACRO EDGE WEAR	VENTRAL	DORSAL
FRACTURES	absent	retouch
FRACTURE TYPE	absent	retouch
ROUNDING	absent	retouch
GLOSS	absent	retouch

MICRO EDGE WEAR

FRACTURES	<5	retouch
FRACTURE TYPE	flakes	retouch
ROUNDING	absent	retouch
MICROTOPOGRAPHY OF POLISHED AREA	flat	retouch
MICRO-POLISH DISTRIBUTION	continuous	retouch
DISTRIBUTION TYPE	edge only even	retouch
INVASIVENESS	<0.5	retouch
LINEAR FEATURES	perpendicular	retouch
STRIATIONS	perpendicular	retouch
POLISH DEVELOPMENT	C	C

BLIND TEST TOOL # 34

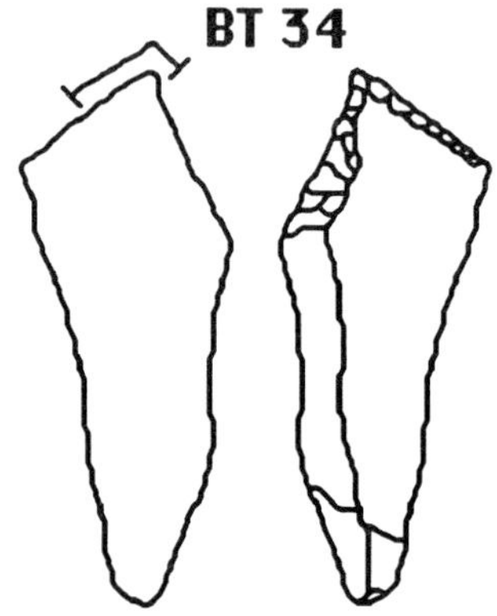

TYPE:	PIERCER
USED AREA:	PIERCER TIP
	morphology, edge wear and polish
DIRECTION OF USE:	LONGITUDINAL
	parallel striations and *away from the edge* distribution indicating a motion along the edge
MOTION OF USE:	GROOVING
	particularly from the *away from the edge* distribution on ventral surface caused by rubbing along the inside of a groove
HARDNESS OF MATERIAL:	MEDIUM TO HARD
	as appropriate to a grooving motion, lack of macro edge wear on such a delicate tip indicates a medium material, light rounding and micro edge wear indicates a *hard* medium material
NAMED MATERIAL:	SOAKED ANTLER
	i.e. a *hard* medium material
PROBABLE FUNCTION:	GROOVING SOAKED ANTLER
ACTUAL USE:	GROOVING SOAKED ANTLER for 7 minutes

OBSERVATIONS

TOOL NO.	34	
TYPE	piercer	
EDGE MORPHOLOGY		
EDGE ANGLE	56	
EDGE LENGTH	25	
THICKNESS	3	
PROFILE	0.67	
SHAPE	0.4	
MACRO EDGE WEAR	VENTRAL	DORSAL
FRACTURES	absent	absent
FRACTURE TYPE	absent	absent
ROUNDING	absent	absent
GLOSS	absent	absent
MICRO EDGE WEAR		
FRACTURES	<5	<5
FRACTURE TYPE	flakes	flakes
ROUNDING	light	light
MICROTOPOGRAPHY OF POLISHED AREA	flat	flat
MICRO-POLISH DISTRIBUTION	continuous	continuous
DISTRIBUTION TYPE	away from edge	away from edge
INVASIVENESS	<0.5	<0.5
LINEAR FEATURES	absent	absent
STRIATIONS	absent	absent
POLISH DEVELOPMENT	C	C

BLIND TEST TOOL # 35

BT 35

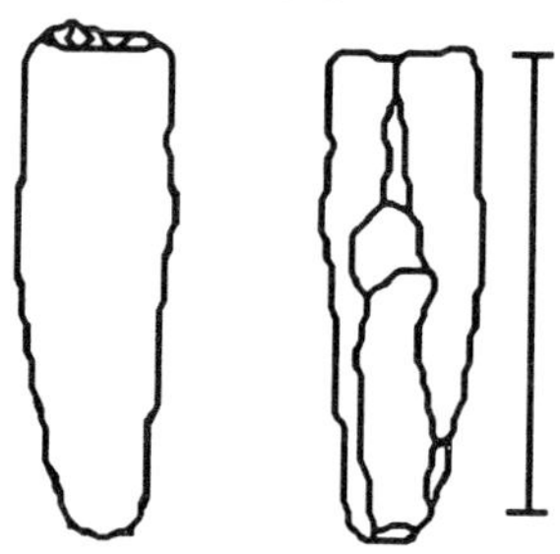

TYPE:	TRUNCATED BLADELET
USED AREA:	RIGHT LATERAL EDGE morphology, edge wear and polish
DIRECTION OF USE:	LONGITUDINAL because of parallel linear features and presence of snap fractures which are often associated with lateral movement of a tool within a cut
MOTION OF USE:	CUTTING as more appropriate than sawing with such a delicate edge (edge angle = 33, thickness = 2)
HARDNESS OF MATERIAL:	SOFT TO MEDIUM soft because of delicate edge, but amount of edge wear indicates a more medium material (macro and micro edge wear = >5) but penetrated material (invasiveness = >0.5D) well developed polish (type C) indicates a yielding material
NAMED MATERIAL:	WOOD/PLANT somewhere between the two
PROBABLE FUNCTION:	CUTTING WOODY PLANT
ACTUAL USE:	CUTTING FRESH SAPLINGS for 18 minutes

OBSERVATIONS

TOOL NO.	35	
TYPE	truncated bladelet	
EDGE MORPHOLOGY		
EDGE ANGLE	33	
EDGE LENGTH	34	
THICKNESS	2	
PROFILE	0.12	
SHAPE	2.8	
MACRO EDGE WEAR	VENTRAL	DORSAL
FRACTURES	<5	<5
FRACTURE TYPE	snaps	snaps
ROUNDING	absent	absent
GLOSS	present	present
MICRO EDGE WEAR		
FRACTURES	>5	<5
FRACTURE TYPE	flakes&snaps	snaps
ROUNDING	light	absent
MICROTOPOGRAPHY OF POLISHED AREA	ridged	flat
MICRO-POLISH DISTRIBUTION	continuous	intermittent
DISTRIBUTION TYPE	edge only even	edge only even
INVASIVENESS	>0.5	>0.5
LINEAR FEATURES	parallel	parallel
STRIATIONS	absent	absent
POLISH DEVELOPMENT	C	C

BLIND TEST TOOL # 36

BT 36

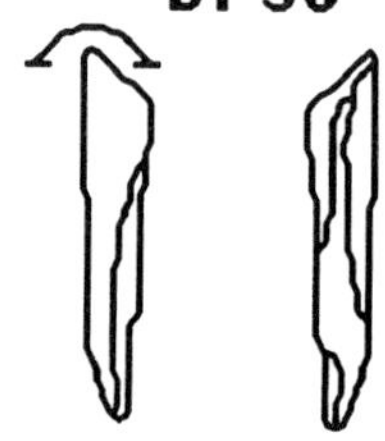

TYPE:	BURIN SPALL
USED AREA:	TIP OF SPALL
	morphology, edge wear and polish
DIRECTION OF USE:	ROTATIONAL
	distribution of polish on the tip, parallel i.e. around the tip striations with a well developed polish (type C) also torsion fractures
MOTION OF USE:	BORING
	from above
HARDNESS OF MATERIAL:	MEDIUM
	hard material would have created rounding, if used on soft material no edge ear would be expected (micro edge wear = >5)
NAMED MATERIAL:	WOOD
PROBABLE FUNCTION:	BORING WOOD
ACTUAL USE:	BORING HORN

For a discussion of horn as a worked material see Chapter 4, Section 10.

OBSERVATIONS

TOOL NO.	36	
TYPE	burin spall [point]	

EDGE MORPHOLOGY

EDGE ANGLE	31
EDGE LENGTH	12
THICKNESS	2
PROFILE	1.2
SHAPE	0.2

MACRO EDGE WEAR	VENTRAL	DORSAL
FRACTURES	<5	<5
FRACTURE TYPE	flakes	flakes
ROUNDING	absent	absent
GLOSS	absent	absent

MICRO EDGE WEAR		
FRACTURES	<5	<5
FRACTURE TYPE	flakes	flakes
ROUNDING	light	light
MICROTOPOGRAPHY OF POLISHED AREA	flat	flat
MICRO-POLISH DISTRIBUTION	continuous	continuous
DISTRIBUTION TYPE	edge only even	edge only even
INVASIVENESS	<0.5	<0.5
LINEAR FEATURES	absent	absent
STRIATIONS	absent	absent
POLISH DEVELOPMENT	C	C

BLIND TEST TOOL #37

BT 37

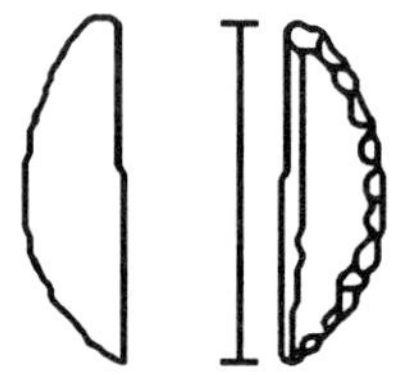

TYPE:	LUNATE MICROLITH
USED AREA:	LEFT LATERAL EDGE
	morphology, edge wear and polish
DIRECTION OF USE:	LONGITUDINAL
	morphology of tool, presence of snap fractures often associated with the bending stress caused by lateral movement of the edge with a longitudinal motion
MOTION OF USE:	CUTTING
	asymmetric polish distribution indicated that the point of the tool was the leading edge followed by a slicing action of the lateral edge, so used at an angle to the worked material therefore not sawing
HARDNESS OF MATERIAL:	SOFT TO MEDIUM
	delicacy of edge (edge angle =34, thickness =2) indicates a soft material but asymmetry of polish distribution indicates the point was in contact with a harder material than with the lateral edge
NAMED MATERIAL:	WOOD/PLANT
	indications of two material, point used on harder (wood) with lateral edge cutting through softer material (plant)
PROBABLE FUNCTION:	CUTTING WOOD
ACTUAL USE:	CUTTING BARK ON A WOODEN BOARD for 14 minutes

OBSERVATIONS

TOOL NO.	37
TYPE	lunate microlith

EDGE MORPHOLOGY

EDGE ANGLE	34
EDGE LENGTH	21
THICKNESS	2
PROFILE	0
SHAPE	2.75

MACRO EDGE WEAR	VENTRAL	DORSAL
FRACTURES	>5	>5
FRACTURE TYPE	snaps	snaps
ROUNDING	absent	absent
GLOSS	present	present

MICRO EDGE WEAR		
FRACTURES	>5	>5
FRACTURE TYPE	snaps	snaps
ROUNDING	absent	absent
MICROTOPOGRAPHY OF POLISHED AREA	ridged	flat
MICRO-POLISH DISTRIBUTION	continuous	continuous
DISTRIBUTION TYPE	edge only asymmetric	edge only asymmetric
INVASIVENESS	<0.5	<0.5
LINEAR FEATURES	parallel	parallel
STRIATIONS	absent	absent
POLISH DEVELOPMENT	C	C

BLIND TEST TOOL #38

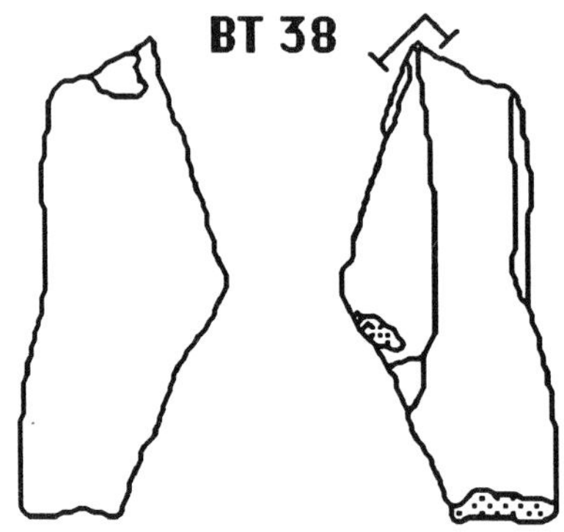

TYPE: TRUNCATION BURIN

USED AREA: BURIN TIP
there was a small amount of polish but this was considered insufficient to indicate use

DIRECTION OF USE:

MOTION OF USE:

HARDNESS OF MATERIAL:

NAMED MATERIAL:

PROBABLE FUNCTION: UNUSED

ACTUAL USE: PIERCING HOLES IN BARK

OBSERVATIONS

TOOL NO.	38	
TYPE	burin bit [point]	
EDGE MORPHOLOGY		
EDGE ANGLE	57	
EDGE LENGTH	15	
THICKNESS	2	
PROFILE	0.64	
SHAPE	0.52	
MACRO EDGE WEAR	VENTRAL	DORSAL
FRACTURES	absent	absent
FRACTURE TYPE	absent	absent
ROUNDING	absent	absent
GLOSS	absent	absent
MICRO EDGE WEAR		
FRACTURES	absent	absent
FRACTURE TYPE	absent	absent
ROUNDING	absent	absent
MICROTOPOGRAPHY OF POLISHED AREA	flat	flat
MICRO-POLISH DISTRIBUTION	absent	absent
DISTRIBUTION TYPE	absent	absent
INVASIVENESS	absent	absent
LINEAR FEATURES	absent	absent
STRIATIONS	absent	absent
POLISH DEVELOPMENT	absent	absent

BLIND TEST TOOL #39

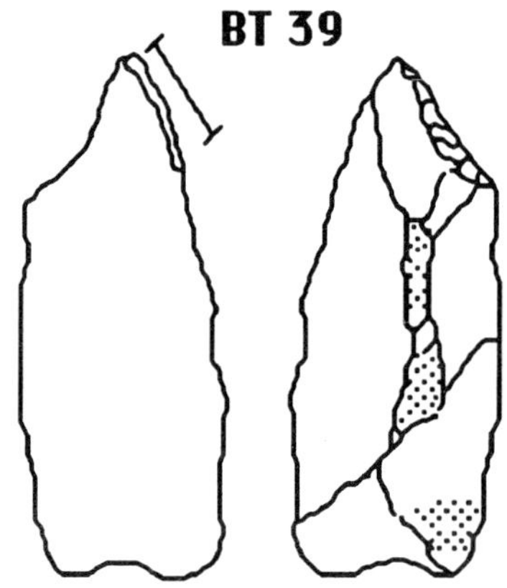

TYPE: TRUNCATION BURIN

USED AREA: BURIN FACET

morphology and polish

DIRECTION OF USE: TRANSVERSE

morphology of edge and angled striations and linear features

MOTION OF USE: SCRAPING

from direction across the edge

HARDNESS OF MATERIAL: HARD

intermittent distribution, linear polish development, general robustness of the tool and the the lack of penetration (invasiveness = <0.5D) indicates a hard material (lack of edge wear due to high edge angle [=68]) linear polish development (type D) indicates a hard resistant material

NAMED MATERIAL: BONE

PROBABLE FUNCTION: SCRAPING BONE

ACTUAL USE: SCRAPING BONE for 7 minutes

OBSERVATIONS

TOOL NO.	39	
TYPE	burin facet	
EDGE MORPHOLOGY		
EDGE ANGLE	68	
EDGE LENGTH	15	
THICKNESS	5	
PROFILE	0.15	
SHAPE	1.44	
MACRO EDGE WEAR	VENTRAL	DORSAL
FRACTURES	absent	absent
FRACTURE TYPE	absent	absent
ROUNDING	absent	absent
GLOSS	absent	absent
MICRO EDGE WEAR		
FRACTURES	absent	absent
FRACTURE TYPE	absent	absent
ROUNDING	absent	absent
MICROTOPOGRAPHY OF POLISHED AREA	flat	flat
MICRO-POLISH DISTRIBUTION	continuous	absent
DISTRIBUTION TYPE	edge only even	absent
INVASIVENESS	<0.5	absent
LINEAR FEATURES	angled	absent
STRIATIONS	angled	absent
POLISH DEVELOPMENT	D	absent

BLIND TEST TOOL # 40

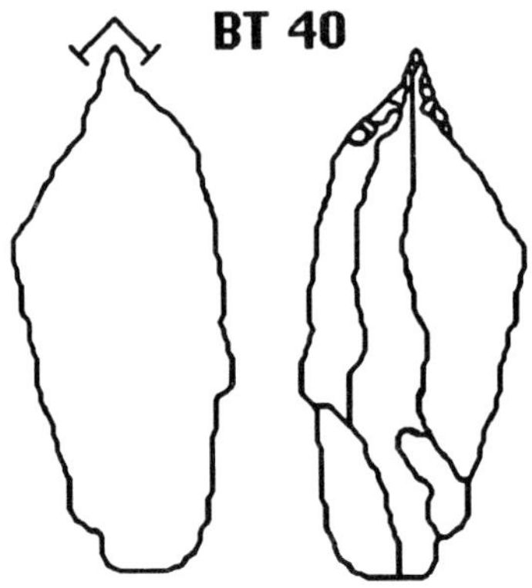

TYPE:	PIERCER
USED AREA:	PIERCER TIP
	morphology, micro edge wear and polish
DIRECTION OF USE:	TRANSVERSE
	perpendicular striations
MOTION OF USE:	PIERCING
	transverse motion with a point
HARDNESS OF MATERIAL:	SOFT
	lack of macro edge wear and small amount of micro edge wear (=<5) on such a delicate tip
NAMED MATERIAL:	HIDE
	soft material suitable for a piecing motion
PROBABLE FUNCTION:	PIERCING HIDE
ACTUAL USE:	PIERCING HIDE for 9 minutes

OBSERVATIONS

TOOL NO.	40	
TYPE	piercer [point]	
EDGE MORPHOLOGY		
EDGE ANGLE	56	
EDGE LENGTH	17	
THICKNESS	3	
PROFILE	0.6	
SHAPE	0.35	
MACRO EDGE WEAR	VENTRAL	DORSAL
FRACTURES	absent	absent
FRACTURE TYPE	absent	absent
ROUNDING	absent	absent
GLOSS	absent	absent
MICRO EDGE WEAR		
FRACTURES	<5	<5
FRACTURE TYPE	flakes	flakes
ROUNDING	absent	absent
MICROTOPOGRAPHY OF POLISHED AREA	flat	flat
MICRO-POLISH DISTRIBUTION	intermittent	intermittent
DISTRIBUTION TYPE	edge only even	edge only even
INVASIVENESS	<0.5	<0.5
LINEAR FEATURES	absent	absent
STRIATIONS	parallel	parallel
POLISH DEVELOPMENT	B	C

<u>**BLIND TEST TOOL # 41**</u>

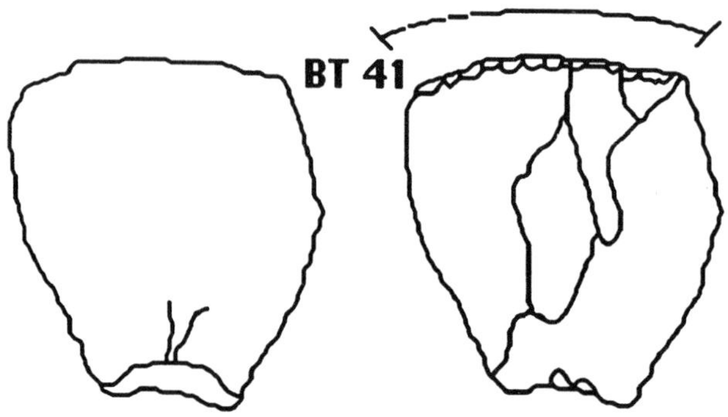

TYPE: ENDSCRAPER

USED AREA: RETOUCHED EDGE
morphology and polish

DIRECTION OF USE: TRANSVERSE
profile (0.87) and edge angle (60) make this edge unsuitable for a longitudinal motion

MOTION OF USE: SCRAPING
only logical motion with the absence of any indications of a percussive motion

HARDNESS OF MATERIAL: SOFT
lack of edge wear indicates soft material

NAMED MATERIAL: HIDE

PROBABLE FUNCTION: SCRAPING HIDE

ACTUAL USE: SCRAPING FRESH WOOD for 30 minutes

Fresh i.e. soft wood was misinterpreted as hide. The softness of the wood accounting for the absence of edge wear.

OBSERVATIONS

TOOL NO.	41
TYPE	endscraper

EDGE MORPHOLOGY

EDGE ANGLE	55
EDGE LENGTH	42
THICKNESS	9
PROFILE	0.33
SHAPE	0.87

MACRO EDGE WEAR	VENTRAL	DORSAL
FRACTURES	absent	absent
FRACTURE TYPE	absent	absent
ROUNDING	absent	absent
GLOSS	absent	absent

MICRO EDGE WEAR

FRACTURES	absent	absent
FRACTURE TYPE	absent	absent
ROUNDING	light	light
MICROTOPOGRAPHY OF POLISHED AREA	flat	flat
MICRO-POLISH DISTRIBUTION	intermittent	absent
DISTRIBUTION TYPE	edge only even	absent
INVASIVENESS	<0.5	absent
LINEAR FEATURES	absent	absent
STRIATIONS	perpendicular	absent
POLISH DEVELOPMENT	B	absent

BLIND TEST TOOL #42

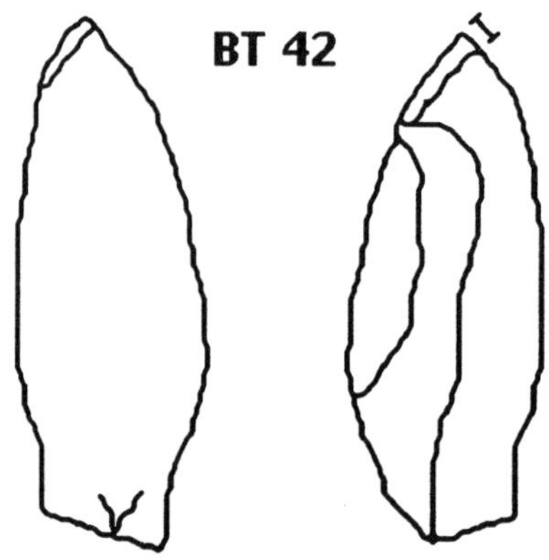

TYPE: DIHEDRAL BURIN

USED AREA: BURIN BIT
morphology, edge wear and polish

DIRECTION OF USE: LONGITUDINAL
polish distributed along the edge and on edges of adjoining facets with edge wear being on sides of bit rather than initiated from pressure across the bit

MOTION OF USE: GROOVING
appropriate to use burin bit for grooving and confirmed by polish distribution

HARDNESS OF MATERIAL: HARD
lack of penetration (invasiveness = edge only) and micro edge wear (<5) on such a robust edge (edge angle =64, length =3, thickness =3, profile =0)

NAMED MATERIAL: STONE
as a 'hard' hard material

PROBABLE FUNCTION: GROOVING STONE

ACTUAL USE: GROOVING SHELL for 7 minutes

The identification of hard material was correct as shell is hard, but there was insufficient information to distinguish between different 'hard' hard materials.

OBSERVATIONS

TOOL NO.	42	
TYPE	dihedral burin bit	
EDGE MORPHOLOGY		
EDGE ANGLE	64	
EDGE LENGTH	3	
THICKNESS	3	
PROFILE	0	
SHAPE	0.4	
MACRO EDGE WEAR	VENTRAL	DORSAL
FRACTURES	absent	absent
FRACTURE TYPE	absent	absent
ROUNDING	absent	absent
GLOSS	absent	absent
MICRO EDGE WEAR		
FRACTURES	<5	<5
FRACTURE TYPE	flakes	flakes
ROUNDING	light	light
MICROTOPOGRAPHY OF POLISHED AREA	flat	flat
MICRO-POLISH DISTRIBUTION	continuous	continuous
DISTRIBUTION TYPE	edge only even	edge only even
INVASIVENESS	edge only	edge only
LINEAR FEATURES	absent	absent
STRIATIONS	absent	absent
POLISH DEVELOPMENT	B+	B+

BLIND TEST TOOL #43

BT 43

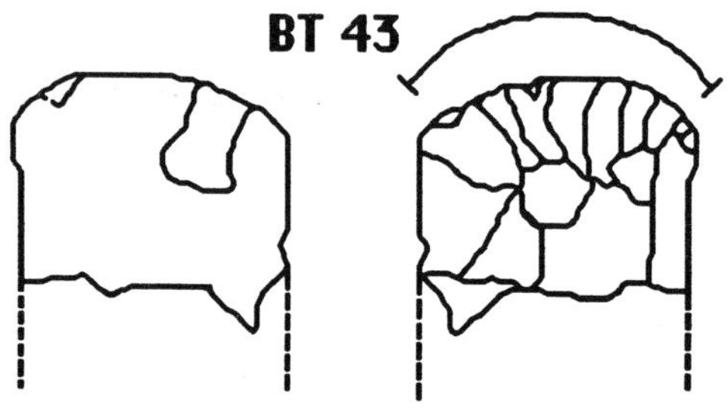

TYPE:	ENDSCRAPER (broken)
USED AREA:	RETOUCHED EDGE
	morphology, edge wear and polish
DIRECTION OF USE:	TRANSVERSE
	perpendicular striation and polish distribution as well as the morphology of the tool
MOTION OF USE:	SCRAPING
	appropriate transverse motion
HARDNESS OF MATERIAL:	MEDIUM
	intermittent polish distribution indicates a less yielding material than say hide, together with lack of penetration (invasiveness = <0.5D) and polish development of B indicates not a hard material
NAMED MATERIAL:	WOOD
PROBABLE FUNCTION:	SCRAPING WOOD

ACTUAL USE: ADZING WOOD for 20 minutes

The motion was misinterpreted because of the failure to diagnose the edge wear (macro step fractures) as resulting from percussion. Also the fact that the tool had broken during during use should have suggested a percussive motion.

OBSERVATIONS

TOOL NO.	43	
TYPE	endscraper [broken]	
EDGE MORPHOLOGY		
EDGE ANGLE	46	
EDGE LENGTH	32	
THICKNESS	5	
PROFILE	0.29	
SHAPE	broken	
MACRO EDGE WEAR	VENTRAL	DORSAL
FRACTURES	<5	retouch
FRACTURE TYPE	steps	retouch
ROUNDING	absent	retouch
GLOSS	absent	retouch
MICRO EDGE WEAR		
FRACTURES	absent	retouch
FRACTURE TYPE	absent	retouch
ROUNDING	absent	retouch
MICROTOPOGRAPHY OF POLISHED AREA	flat	retouch
MICRO-POLISH DISTRIBUTION	intermittent	retouch
DISTRIBUTION TYPE	edge only even	retouch
INVASIVENESS	<0.5	retouch
LINEAR FEATURES	absent	retouch
STRIATIONS	perpendicular	retouch
POLISH DEVELOPMENT	B	retouch

BLIND TEST TOOL #44

BT 44

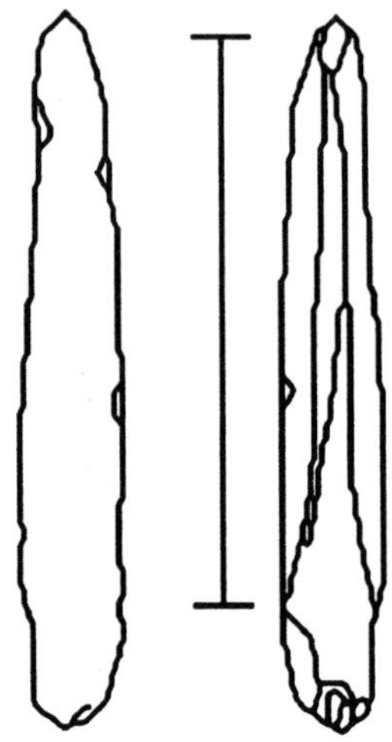

TYPE: BLADE

USED AREA: RIGHT LATERAL EDGE
morphology, edge wear and polish

DIRECTION OF USE: LONGITUDINAL
morphology of edge (angle =38, thickness =4, profile =0.04) and snap fractures associated with bending stress from lateral movement of edge in a longitudinal motion

MOTION OF USE: CUTTING/SAWING
from above

HARDNESS OF MATERIAL: MEDIUM/HARD
morphology of tool mitigates against a hard material, but rounding and edge wear indicates hard, also lack of penetration (invasiveness = <0.5D) so on balance hard

NAMED MATERIAL: BONE/ANTLER

PROBABLE FUNCTION: CUTTING BONE

ACTUAL USE: CUTTING CORTEX ON FLINT NODULE for 11 minutes

This was 'recognised' as an *unknown material* (see Chapter 4, Section 10)

OBSERVATIONS

TOOL NO.	44	
TYPE	blade	
EDGE MORPHOLOGY		
EDGE ANGLE	38	
EDGE LENGTH	38	
THICKNESS	4	
PROFILE	0.04	
SHAPE	4.57	
MACRO EDGE WEAR	VENTRAL	DORSAL
FRACTURES	>5	>5
FRACTURE TYPE	snaps	snaps
ROUNDING	light	light
GLOSS	present	present
MICRO EDGE WEAR		
FRACTURES	>5	>5
FRACTURE TYPE	flakes&snaps	flakes&snaps
ROUNDING	heavy	heavy
MICROTOPOGRAPHY OF POLISHED AREA	flat	flat
MICRO-POLISH DISTRIBUTION	continuous	continuous
DISTRIBUTION TYPE	edge only even	edge only even
INVASIVENESS	<0.5	<0.5
LINEAR FEATURES	absent	absent
STRIATIONS	parallel	parallel
POLISH DEVELOPMENT	C	C

BLIND TEST TOOL #45

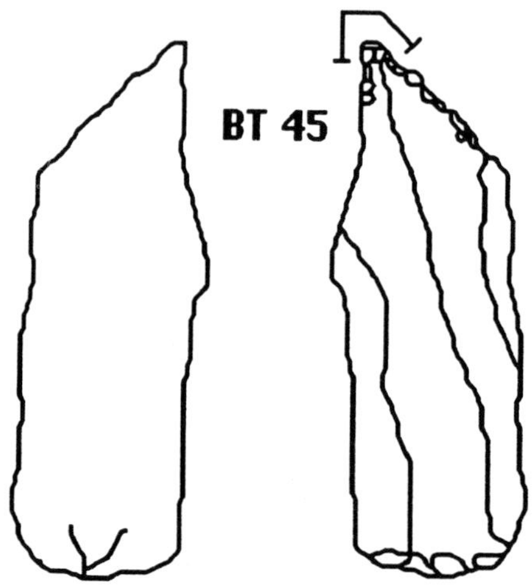

TYPE:	PIERCER
USED AREA:	PIERCER TIP (broken) morphology of tool and polish
DIRECTION OF USE:	TRANSVERSE i.e. to the apex of the point from perpendicular linear features and striations
MOTION OF USE:	GROOVING transverse motion with asymmetric polish distribution indicates one side of the tip was the leading edge so an *off-centre* working orientation like grooving rather than piercing
HARDNESS OF MATERIAL:	MEDIUM grooving action indicates at least a medium material and lack of edge wear indicates not a hard material, but the presence of linear features and intermittent polish distribution indicates a 'hard' medium material
NAMED MATERIAL:	WOOD/SOAKED ANTLER soaked antler more probable because of indications that worked material was less yielding than wood
PROBABLE FUNCTION:	GROOVING SOAKED ANTLER
ACTUAL USE:	GROOVING SOAKED ANTLER for 15 minutes

OBSERVATIONS

TOOL NO.	45	
TYPE	piercer	
EDGE MORPHOLOGY		
EDGE ANGLE	35	
EDGE LENGTH	6	
THICKNESS	11	
PROFILE	0.66	
SHAPE	0.74	
MACRO EDGE WEAR	VENTRAL	DORSAL
FRACTURES	absent	retouch
FRACTURE TYPE	absent	retouch
ROUNDING	absent	retouch
GLOSS	absent	retouch
MICRO EDGE WEAR		
FRACTURES	absent	retouch
FRACTURE TYPE	absent	retouch
ROUNDING	absent	retouch
MICROTOPOGRAPHY OF POLISHED AREA	flat	retouch
MICRO-POLISH DISTRIBUTION	intermittent	retouch
DISTRIBUTION TYPE	edge only asymmetric	retouch
INVASIVENESS	<0.5	retouch
LINEAR FEATURES	perpendicular	retouch
STRIATIONS	perpendicular	retouch
POLISH DEVELOPMENT	B+	retouch

BLIND TEST TOOL #46

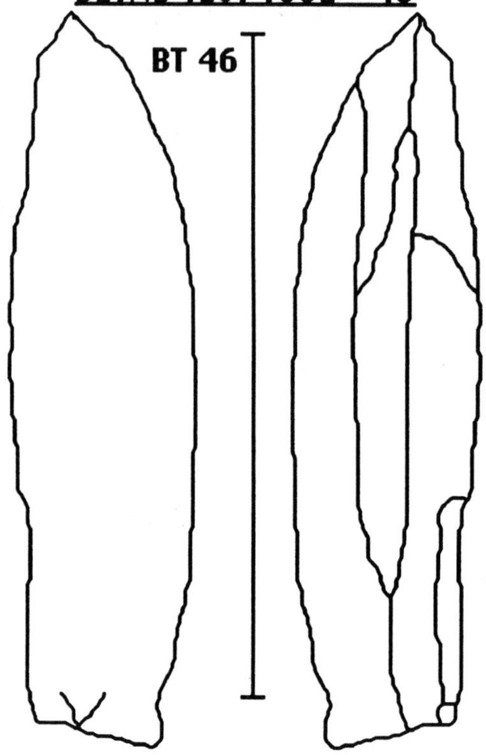

TYPE:	BLADE
USED AREA:	LEFT LATERAL EDGE
	morphology, edge wear and polish
DIRECTION OF USE:	LONGITUDINAL
	morphology of edge (angle =26, length =94, thickness =6) and presence of snap fractures
MOTION OF USE:	CUTTING
	edge is too delicate for sawing
HARDNESS OF MATERIAL:	SOFT/MEDIUM
	contact with a hard material would have produced more edge wear, but macro edge wear (>5) indicates a medium rather than soft material
NAMED MATERIAL:	WOOD/PLANTS
	edge wear is consistent with a medium yielding material as there are indications of both soft and medium materials
PROBABLE FUNCTION:	CUTTING WOODY PLANT

ACTUAL USE: CUTTING FISH for 8 minutes

Identification of a 'soft' medium material was correct but there was no evidence to specifically identify fish

OBSERVATIONS

TOOL NO.	46	
TYPE	blade	
EDGE MORPHOLOGY		
EDGE ANGLE	26	
EDGE LENGTH	94	
THICKNESS	6	
PROFILE	0.1	
SHAPE	3.9	
MACRO EDGE WEAR	VENTRAL	DORSAL
FRACTURES	>5	>5
FRACTURE TYPE	snaps	snaps
ROUNDING	absent	retouch
GLOSS	absent	retouch
MICRO EDGE WEAR		
FRACTURES	absent	absent
FRACTURE TYPE	absent	absent
ROUNDING	absent	absent
MICROTOPOGRAPHY OF POLISHED AREA	ridged	flat
MICRO-POLISH DISTRIBUTION	intermittent	continuous
DISTRIBUTION TYPE	edge only even	edge only even
INVASIVENESS	>0.5	<0.5
LINEAR FEATURES	absent	absent
STRIATIONS	absent	absent
POLISH DEVELOPMENT	B+	B

BLIND TEST TOOL #47

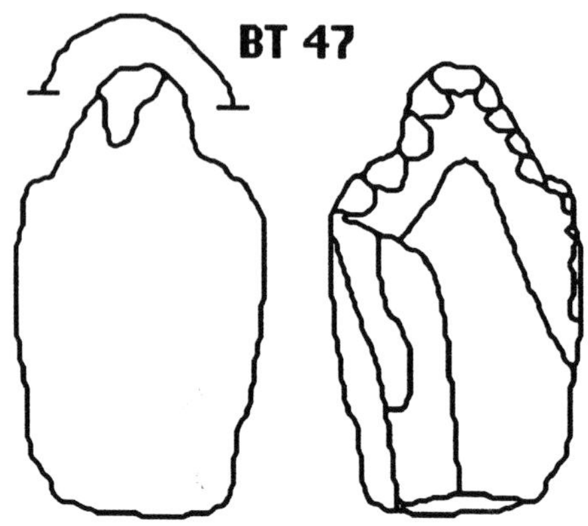

TYPE:	ENDSCRAPER
USED AREA:	RETOUCHED EDGE presence of polish
DIRECTION OF USE:	TRANSVERSE too thick and too convex for a cutting or sawing action motion
MOTION OF USE:	SCRAPING as no indication of percussion
HARDNESS OF MATERIAL:	MEDIUM lack of edge wear, hard material may have produced a more developed polish whereas this was B+
NAMED MATERIAL:	WOOD/SOAKED ANTLER lack of polish development indicates softer of these two materials
PROBABLE FUNCTION:	SCRAPING WOOD and then trampled -from isolated spots of polish and random orientated striation over both surfaces
ACTUAL USE:	SCRAPING WOOD for 6 minutes and then trampled for 3 minutes

OBSERVATIONS

TOOL NO.	47	
TYPE	shouldered endscraper	
EDGE MORPHOLOGY		
EDGE ANGLE	52	
EDGE LENGTH	30	
THICKNESS	10	
PROFILE	0.8	
SHAPE	0.48	
MACRO EDGE WEAR	VENTRAL	DORSAL
FRACTURES	absent	retouch
FRACTURE TYPE	absent	retouch
ROUNDING	absent	retouch
GLOSS	absent	retouch
MICRO EDGE WEAR		
FRACTURES	absent	retouch
FRACTURE TYPE	absent	retouch
ROUNDING	light	light
MICROTOPOGRAPHY OF POLISHED AREA	flat	retouch
MICRO-POLISH DISTRIBUTION	continuous	retouch
DISTRIBUTION TYPE	edge only even	retouch
INVASIVENESS	<0.5	retouch
LINEAR FEATURES	absent	absent
STRIATIONS	absent	absent
POLISH DEVELOPMENT	B+	retouch

BLIND TEST TOOL #48

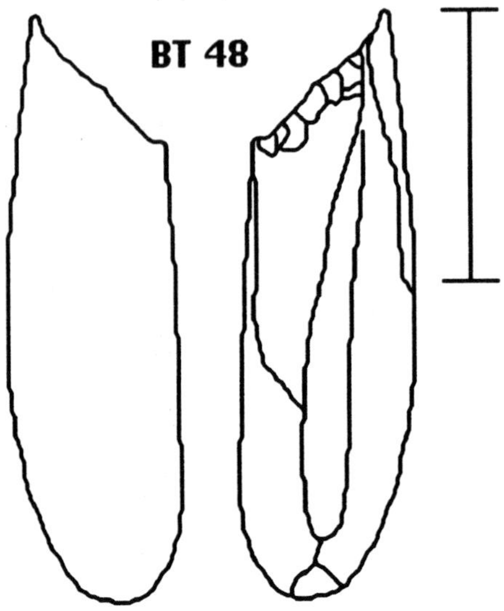

TYPE: TRUNCATION BURIN

USED AREA: BURIN FACET

morphology, edge wear and polish

DIRECTION OF USE: TRANSVERSE

as polish is along most of the length of the facet edge and not concentrated on the bit, micro edge wear along facet from transverse motion in order to produce sufficient pressure to wear such a robust edge (angle =73)

MOTION OF USE: SCRAPING

from morphology of edge as there is no sharp cutting edge and no point, transverse motion rules out grooving so scraping

HARDNESS OF MATERIAL: HARD

morphology of edge, and polish being limited to edge indicates little penetration, polish development (C) indicates hard material from short period of use (or more edge wear or rounding would be expected)

NAMED MATERIAL: BONE/ANTLER

micro edge wear (particularly the step fractures) indicates the harder material

PROBABLE FUNCTION: SCRAPING BONE

ACTUAL USE: SCRAPING BONE for 6 minutes

OBSERVATIONS

TOOL NO.	48	
TYPE	truncation burin	

EDGE MORPHOLOGY		
EDGE ANGLE	81	
EDGE LENGTH	25	
THICKNESS	5	
PROFILE	0	
SHAPE	3.9	

MACRO EDGE WEAR	VENTRAL	DORSAL
FRACTURES	absent	absent
FRACTURE TYPE	absent	absent
ROUNDING	absent	absent
GLOSS	absent	absent

MICRO EDGE WEAR		
FRACTURES	>5	absent
FRACTURE TYPE	flakes&steps	absent
ROUNDING	absent	absent
MICROTOPOGRAPHY OF POLISHED AREA	flat	flat
MICRO-POLISH DISTRIBUTION	continuous	continuous
DISTRIBUTION TYPE	edge only even	edge only even
INVASIVENESS	edge only	<0.5
LINEAR FEATURES	absent	absent
STRIATIONS	absent	absent
POLISH DEVELOPMENT	C	C

<u>**BLIND TEST TOOL #49**</u>

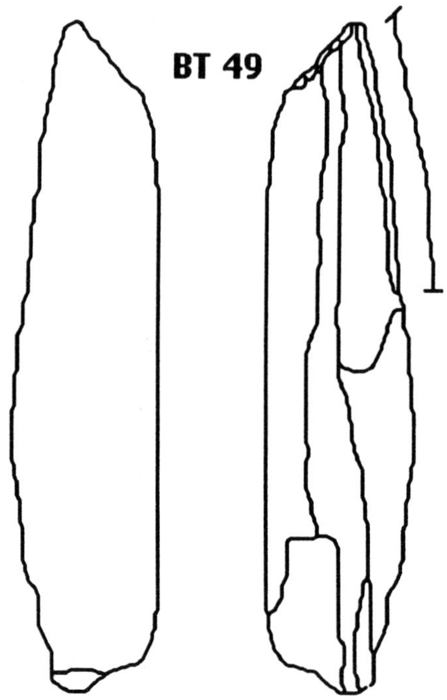

TYPE: TRUNCATION BURIN

USED AREA: BURIN FACET
morphology, edge wear and polish

DIRECTION OF USE: TRANSVERSE
facet has no cutting edge

MOTION OF USE: SCRAPING
dorsal facet surface has perpendicular linear features and striations, with some angled striations that indicate a variation in the direction of motion consistent with scraping with a facet edge

HARDNESS OF MATERIAL: MEDIUM/HARD
presence of light rounding indicates a hard material but lack of edge wear indicates medium so a 'hard' medium material

NAMED MATERIAL: SOAKED ANTLER
from above

PROBABLE FUNCTION: SCRAPING SOAKED ANTLER

ACTUAL USE: SCRAPING SOAKED ANTLER for 25-30 minutes

OBSERVATIONS

TOOL NO.	49	
TYPE	truncation burin facet	
EDGE MORPHOLOGY		
EDGE ANGLE	79	
EDGE LENGTH	29	
THICKNESS	5	
PROFILE	0	
SHAPE	3.95	
MACRO EDGE WEAR	VENTRAL	DORSAL
FRACTURES	absent	absent
FRACTURE TYPE	absent	absent
ROUNDING	absent	absent
GLOSS	absent	absent
MICRO EDGE WEAR		
FRACTURES	absent	absent
FRACTURE TYPE	absent	absent
ROUNDING	light	light
MICROTOPOGRAPHY OF POLISHED AREA	flat	flat
MICRO-POLISH DISTRIBUTION	continuous	continuous
DISTRIBUTION TYPE	edge only even	edge only even
INVASIVENESS	<0.5	<0.5
LINEAR FEATURES	absent	perpendicular
STRIATIONS	absent	perpendicular
POLISH DEVELOPMENT	C	C

BLIND TEST TOOL #50

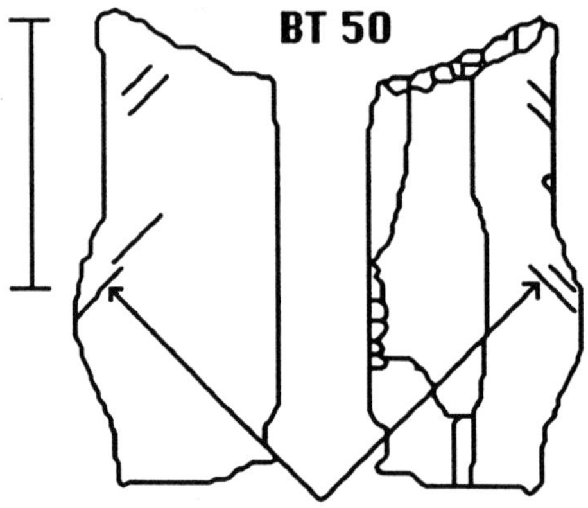

TYPE:	TRUNCATED BLADE
USED AREA:	LEFT LATERAL EDGE
	morphology of tool, backing retouch on right edge presence of edge wear and polish
DIRECTION OF USE:	TRANSVERSE
	orientation of linear features
MOTION OF USE:	CUTTING/WHITTLING
	has a thin edge with edge wear mainly of snap fractures, whittling is more probable because of angled linear features and polish being more invasive on one surface
HARDNESS OF MATERIAL:	MEDIUM
	edge wear indicates not a soft material and to create such linear features requires at least a medium material, invasiveness of polish indicates penetration so a yielding medium material
NAMED MATERIAL:	WOOD
	wood is most probable from above, but the lack of polish development possibly indicates a slightly softer material such as a woody plant like bark
PROBABLE FUNCTION:	WHITTLING WOOD and hafted

ACTUAL USE: WHITTLING WOOD for 17 minutes (hafted)

See Chapter 4, Section 8 for a detailed discussion of this tool.

OBSERVATIONS

TOOL NO.	50	
TYPE	truncated blade	
EDGE MORPHOLOGY		
EDGE ANGLE	40	
EDGE LENGTH	45	
THICKNESS	5	
PROFILE	0.06	
SHAPE	2.24	
MACRO EDGE WEAR	VENTRAL	DORSAL
FRACTURES	>5	>5
FRACTURE TYPE	flakes&snaps	flakes&snaps
ROUNDING	absent	absent
GLOSS	absent	absent
MICRO EDGE WEAR		
FRACTURES	>5	>5
FRACTURE TYPE	flakes&snaps	flakes&snaps
ROUNDING	absent	absent
MICROTOPOGRAPHY OF POLISHED AREA	flat	flat
MICRO-POLISH DISTRIBUTION	continuous	continuous
DISTRIBUTION TYPE	edge only even	edge only even
INVASIVENESS	>0.5	<0.5
LINEAR FEATURES	angled	angled
STRIATIONS	absent	absent
POLISH DEVELOPMENT	B+	B+

APPENDIX 4: THE FAST PROGRAM

This computer program is an expert system program developed using the Hypercard™ application on an Apple Macintosh™ computer. The Hypercard application employs **cards** which have **fields** in which data and information can be entered, and **buttons** which activate programs. The program was written in Hypertalk™ which is the programming language for Hypercard.

The first stage in the development of an expert system is to design the data base for storing the information in such a way that it can be easily accessed and input into the program. The Hypercard application is ideal for this as it's design is based on the idea of a card index file (hence Hypercard). The data card designed for this purpose is illustrated in Figure 92. The observations are entered into the appropriate fields and another stack is created to contain drawings of the flint and each data card is linked to the pertinent drawing. The data for blind test tools 1-10 is illustrated on these data cards in Figure 92, and drawings of the tools are illustrated in Appendix 2.

The data is transferred into the **FAST** (Functional Analysis of Stone Tools) program. Each attribute, that is the value of each variable (e.g. edge angle of 50 degrees), is used to give an indication of motion, or hardness of material, or both, according to a set of rules. For example the variable edge angle is divided into ranges so that the value of the edge angle for a particular tool will fall within one of those ranges giving the corresponding indication i.e. if the edge angle = 50 degrees this indicates `cutting or scraping a medium material`. This process is repeated for each attribute. The program automatically assess the attributes and enters the relevant indications into two cards, one of which contains information concerning macro observations and the other concerning micro observations. Then the incidence of particular motions or materials are counted according to another set of rules. For example each variable counts as two points, one each for the dorsal and ventral surfaces, if the indication is of two alternative materials such as `soft to medium` then each is counted as half a point. Certain motions count double. For example whittling is counted as double because cutting is used as a general term for longitudinal cutting motions, whereas whittling is a more precise motion i.e. cutting at an angle to the tool (see Chapter 4 Section 7.4 and Figure 67 for a definition of whittling). When the indications have been counted according to the rules the scores for motions and materials are entered into the interpretation card. The program is then activated which assesses the numbers of indications for motions and hardness of materials and makes an interpretation of the function of the tool in three categories. The first is motion, for which it selects the most probable motion given the numbers and relationship of the indications of motions. Then the hardness category of the worked material is assigned ranging from soft through soft/medium, medium, medium/hard to hard. The third interpretation is of the named worked material hide, wood, bone etc.

Prior to the development of the expert system computer program (hereafter referred to as **FAST**) the interpretation of each attribute had to be done by assessing this information and the complex interrelationships between attributes in ones head, as it where. **FAST** carries out this process automatically.This not only speeds up the process but makes it completely consistent as the same set of rules are applied each time. The 20 tools used in the last blind test carried out at the Institute (see Chapter 4 Section 10, and Appendix 3) were used to determine the parameters by which the rules were applied in order to make the functional interpretations. That is, the data that was recorded for that blind test was used as the training data for developing the program. The efficacy of the program is demonstrated by it achieving a result of **18** out of **20** correct interpretations of named materials. The two tools that were not correctly identified were tool 38 for which the computer gave "insufficient data" (used on bark) and tool 44 which was designated "unknown" (this was used on cortex which was not programmed into the computer). The same scoring system as used in the blind test was applied, therefore to achieve a point the precise worked material had to be identified. If the tool was used on antler than only an answer of antler was awarded a point, not alternatives like bone/antler. This 90% success rate is a significant increase on the result achieved by any of the analysts in the blind test, the maximum score achieved being 55%. However, as the blind test data was used to develop the program this high rate of success is misleading. The real test of the program is when a completely new set of data is used. The first 10 tools used in blind tests at the Institute were observed and the data recorded as set out in Chapter 4, and then used to test the program. This blind test was reported in Newcomer et al 1987. Of the 10 test tools, one was unused and another was used as a projectile point that struck unknown material, leaving eight precisely known materials that the tools were used on. Of these **8** tools, the precise worked material on which **6** of the tools had been used was identified by the program. The two not identified were tool 2 for which the computer gave "unknown" (actually used on shell), and tool 10 for which the computer gave "antler" but which was used on wood. In the original test only two of these eight were correctly identified. Though the function of these tools was known, this was a blind test as the computer did not have this information.

A necessary aspect of the program is that it has to be adaptable to different archaeological assemblages. The blind test tools used to generate the parameters for this program would fit within an Upper Palaeolithic Magdelanian assemblage, and certain of the rules used in the program include information about burins and whether the used edge of burins was the bit, or the facet. Therefore a program that was developed from that set of tools would not be applicable to, for example, tools from a Mousterian assemblage. Any microwear analysis requires the generation of a reference collection of experimental tools that replicates the archaeological assemblage to be analysed. Microwear analysis involves comparing archaeological tools with this reference collection whether using simple optical microscopy or an expert computer system. Therefore if a set of Mousterian tools is to be analysed then replicas of Mousterian tools used in simulation experiments on materials likely to have been used on the site (based on the environmental and artifactual evidence) would be created and the parameters for the program generated from this experimental collection. To allow for this required adaptability the program has been written so that it can easily be changed to accommodate different archaeological assemblages. A card is designed for each variable so that the analyst can assess each variable individually and change the parameters of the variable ranges along with the relevant functional indications to fit the archaeological assemblage that is to be analysed. When the parameters have been changed so that a *goodness of fit* has been achieved, that is that the parameters are achieving a high success rate on the experimental reference collection (in effect the computer program is blind tested) then the rules for each variable can be combined into one composite program. This means that only a single button is pressed to automatically carry out the program on the variables, rather than going through each variable individually (see Figure 94).

EXAMPLE OF FAST ;TOOL 33

On the data card (see Figure 92) there is a button for each variable which when pressed puts the value of that variable into the appropriate cards. For example, edge angle is put into the **edge angle** card and also into the **gloss, micro-fracture, micro-distribution type** cards etc. **Length**, in addition to being placed into the **length** card, is also placed into the **polish distribution** card. Tool type is put into the **shape** card because one of the diagnostic features depends on whether the tool is a burin or not, so that it is necessary to have that information on the **shape** card. Then, as each variable card is accessed the range within which the value falls is identified and pressing that field enters the appropriate indication into the diagnostic cards. In the example given the relevant indications are printed in italics. In the **edge angle** card the value of 42 degrees for tool 33 falls within the 30-55 degree range, therefore the indication is *cutting or scraping medium material.*. In the **micro-fracture** variable card the value of <5 fractures per 5mm. of working edge (with an edge angle of 42 degrees), gives the indication of *soft to medium material* for the ventral surface. The dorsal surface is retouched and therefore *retouch* is placed in the diagnostic card for the amount of micro-fractures on the dorsal surface. Certain attributes are not diagnostic. For example, for **micro-fracture type** the attribute flakes is not diagnostic and therefore *non diagnostic* is entered for the ventral surface with the indication for the dorsal surface again being *retouch.* The indications are entered into the diagnostic cards and then added together according to a set of rules. These state that each variable has a diagnostic value of 2, except for **thickness** which carries a maximum of 1 point. This is because thickness is not a very diagnostic attribute, there being only a simple division between 1-4 and >4 mm. Morphological attributes such **angle, length, shape** etc., have a value of 2 points because they include information about both the ventral and dorsal surfaces. The other variables have a diagnostic value of 1 point for the ventral and 1 point for the dorsal surface. Where there are alternatives of more than one material or motion within one diagnostic field the alternatives are valued at 0.5 points, unless the other surface is designated *retouch, no polish or no effect.,* in which case they carry a value of 1 point. This means that from the macro diagnostic card for tool 33 there is 1 indication of *cutting* from **edge angle,** 8 indications of *scraping* ; 1 from **edge angle,** 2 from **length,** 1 from **thickness,** 2 from **profile** and 2 from **shape.** There is 1 indication of *soft material* from **rounding** and 5 indications of *medium material* ; 2 from **edge angle,** 2 from amount of **fractures** and 1 from **rounding.** From the micro diagnostic card there are 2 indications of *scraping* from **linear features and striations,** 4 indications of *soft material* ; 1 from **fractures,** 1 from **rounding** and 2 from **polish development.** The rules state that when there are alternatives such as *soft or short time[hard]* the majority from the other variables is counted. Therefore in this case **polish development** indicates the tool was used on a *soft material* rather than on a *hard material* for a short time. There are 3 indications of *medium* ; 1 from **fractures,** 1 from **rounding** and 1 from **invasiveness.** There is 1 indication of *hard* material from **invasiveness.** These indications are totalled on the interpretation card giving 1 indication of *cutting,* 10 indications of *scraping,* 5 indications of *soft* material, 8 indications of *medium* material and 1 indication of *hard* material. Given the parameters of the function program this combination leads to the interpretation that tool 33 was used for SCRAPING a SOFT/MEDIUM material, most probable HIDE.

The part of the Function program that covers this tool and lead to the interpretation of scraping hide are as follows:

if CUT ≤4 and SCRAPE >8 and GROOVE ≤2 and WHITTLING ≤2 then motion = **SCRAPING**

if SOFT ≥4 but ≤12 and MEDIUM >2 but ≤9 and HARD ≥0 but ≤2 then material = **SOFT/MEDIUM**

if SOFT ≥2 but ≤6 and MEDIUM ≤8 and HARD <2 and motion is not WHITTLING or BORING/DRILLING or GROOVING or CHOPPING/ADZING then named material is probably **HIDE**

There is a separate program for POINTS as points are measured in different ways because the whole morphology of the tool is integral to its function. Also certain variables have different values, as measured for points, and therefore different indications. For example the indication card for **fracture types** for the POINT program is illustrated below.

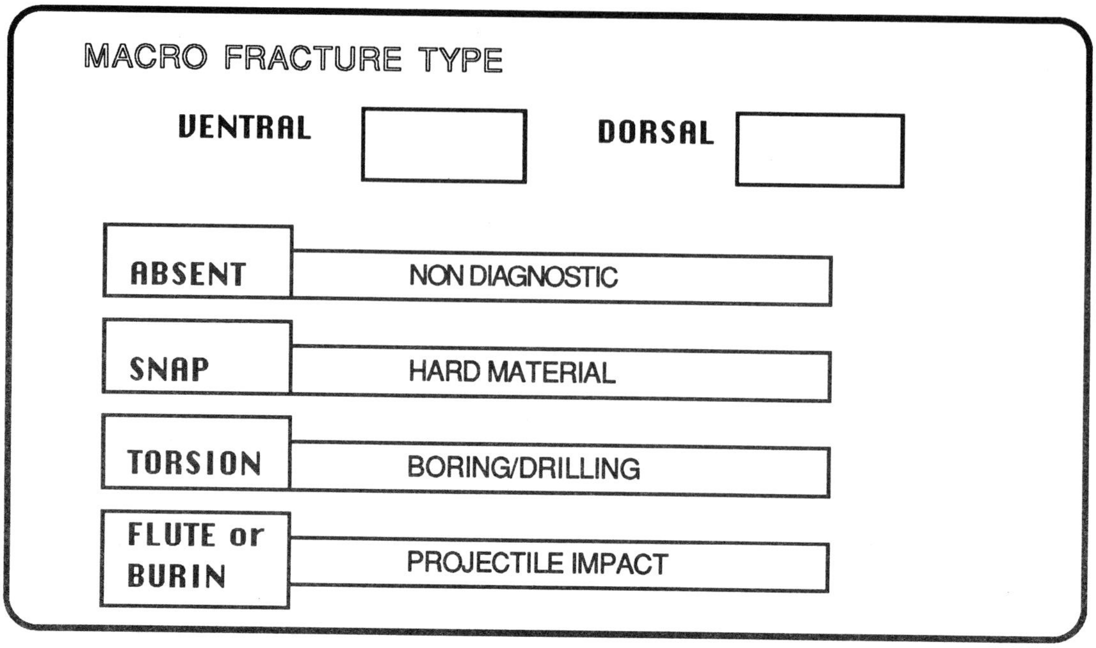

The values of the variable, **macro fracture type**, for the POINT program are absent, snap, torsion and flute or burin. Snap fractures occur when the tip of the point is "snapped off", rather than snap fractures occurring on an edge. For an illustration of torsion, flute and burin fractures see Chapter 5, Figure 89. The indication from these types of "point" fractures can be seen from the card above. Torsion fractures can occur on projectiles, and flute or burin fractures can occur with rotational motions such as drilling, but the indications are not meant to be mutually exclusive but represent the most probable association.

In the POINT program some variables have the same values and the same ranges that determine the indications but the specific indication will be different. For example, edge thickness is measured in mm. and the ranges are 1-4 and >4, but whereas 1-4 indicates cutting in the non point program it indicates a piercing,boring or drilling motion in the POINT program, and >4 with non pointed tools indicates scraping but for points it indicates grooving. This is because the thicker points are stronger and therefore more suitable for grooving than more delicate points that are more suitable for piercing, boring or drilling because the intention with these motions is to penetrate the material rather than gouge out a groove.

As with any complex computer program it is more difficult to describe how it works than to actually use it. Once the parameters of the variables and the relevant indications have been established the operation of the program is extremely simple. As mentioned above the individual variable cards can be dispensed with and incorporated into a program that can be activated from one card. This involves pushing a few buttons and making certain decisions based on a set of rules so that the interpretation of the function of a tool can be carried out within a few minutes (see Figure94). The use of **FAST** not only makes the interpretations consistent but also speeds up the process so that more tools can be processed thus increasing the size of the sample of tools that can be analysed.

PLATE 1

a) unused flint surface [200x]

b) unused flint surface [200x]

PLATE 2

surface of tool used for whittling wood

a) ventral surface

b) dorsal surface

PLATE 3

test tool 8 used for cutting meat

a) unused area

b) used area

<u>PLATE 4</u>

test tool 4 used for scraping stone [200x]

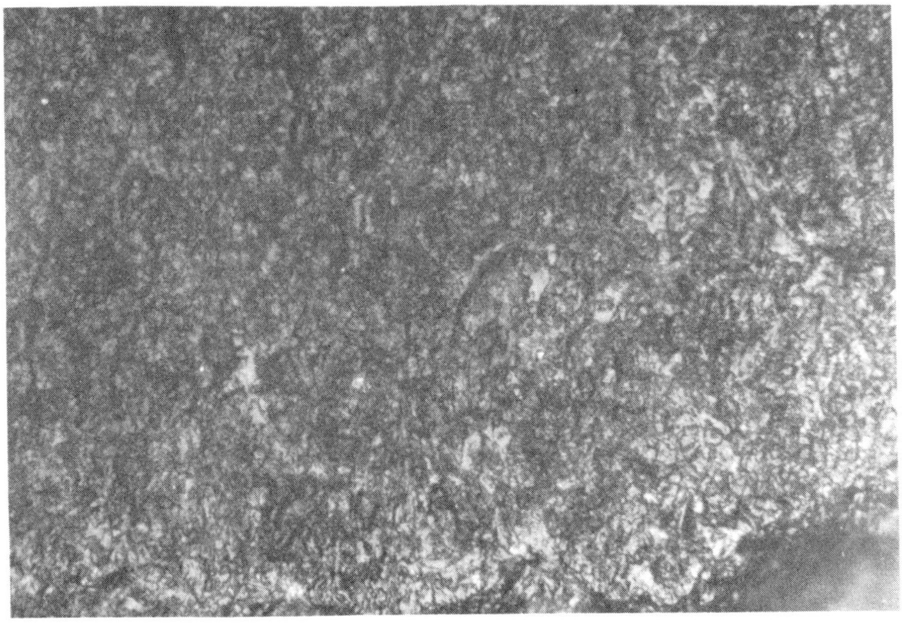

PLATE 5

a) test tool 9 used for scraping fish [200x]

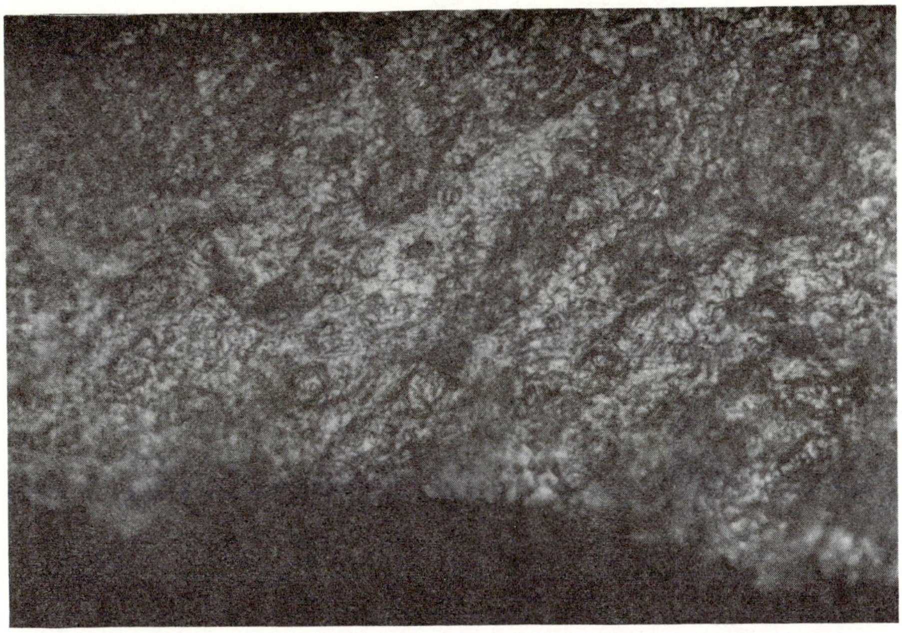

b) test tool 10 used for scraping wood [200x]

<u>PLATE 6</u>

a) polish on dorsal ridge of drill bit before hafting [200x]

b) polish on dorsal ridge of drill bit after hafting [200x]

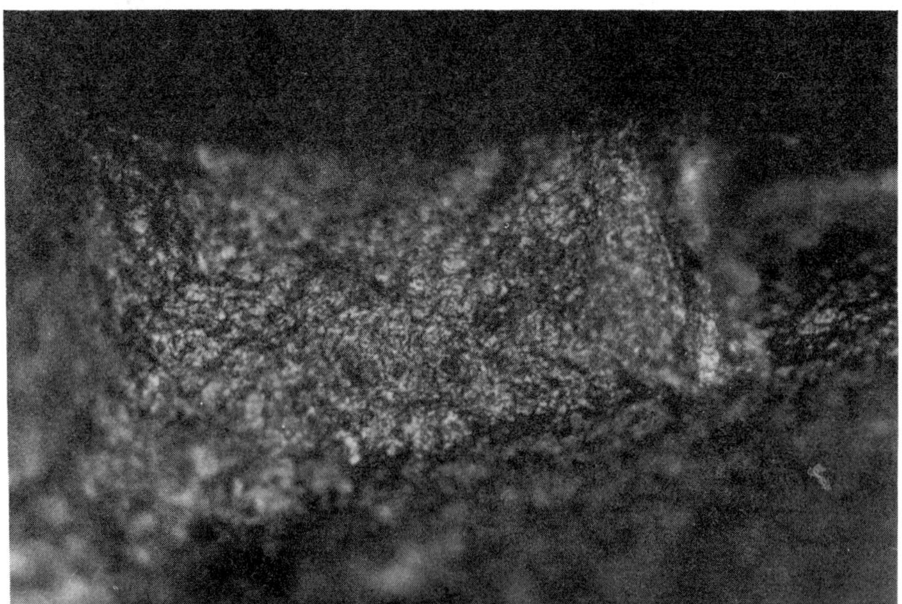

PLATE 7

away from the edge polish distribution on test tool 34 [200x]

PLATE 8

linear feature on test tool 50 [200x]

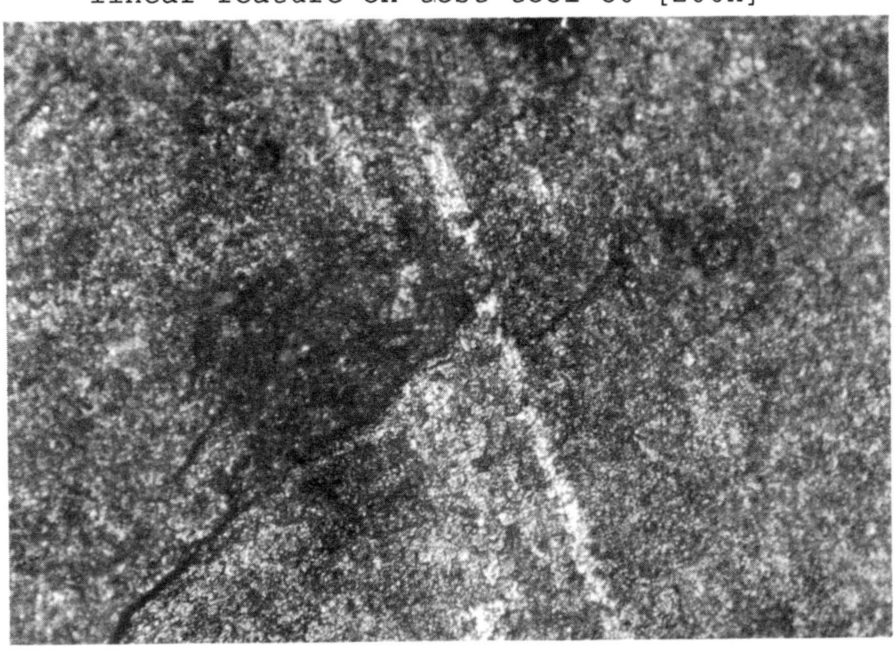

PLATE 9

polish and rounding on test tool 44 used for cutting cortex [200x]

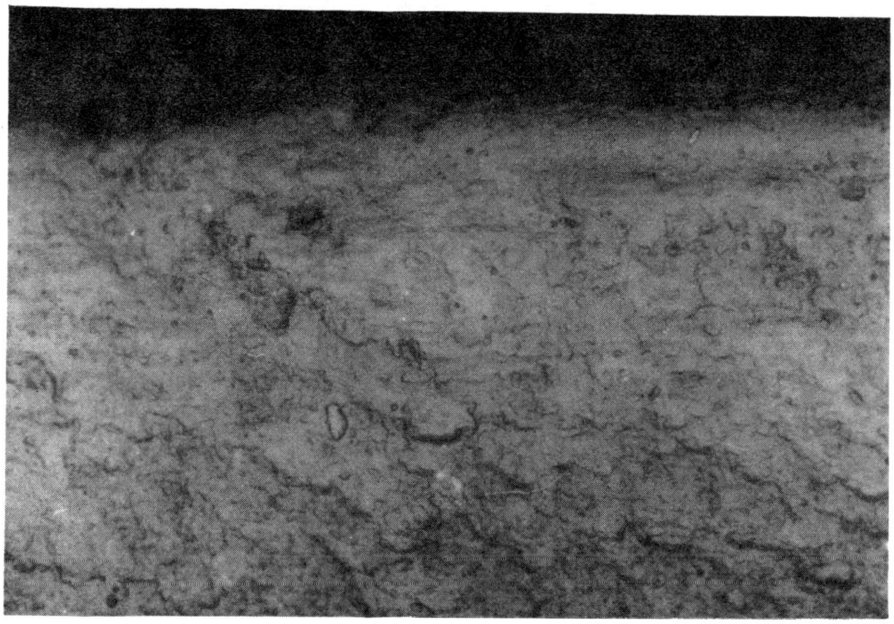

PLATE 10

drill bit from Kumartepe illustrating the rounding at the tip (scale= 3 x actual size)

PLATE 11

tips of experimental drill bits used on carnelian

a [100x]

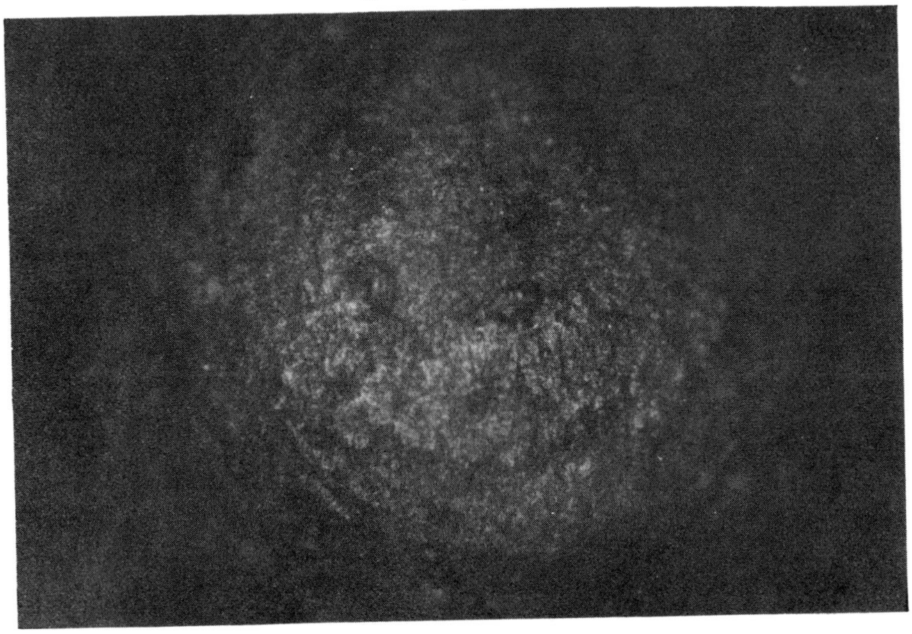

b [100x]

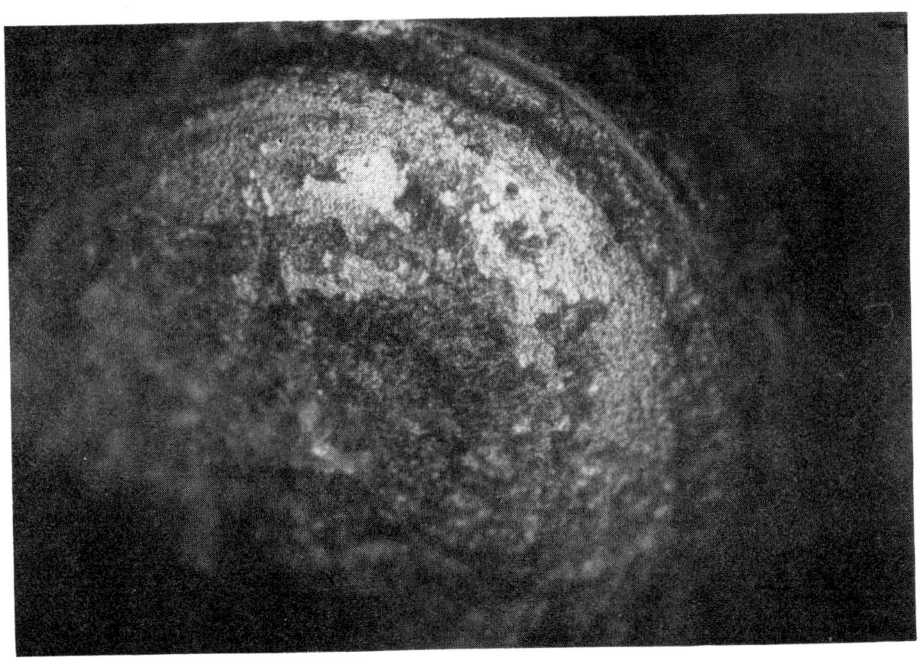

PLATE 12

tips of archaeological drill bits

a [100x]

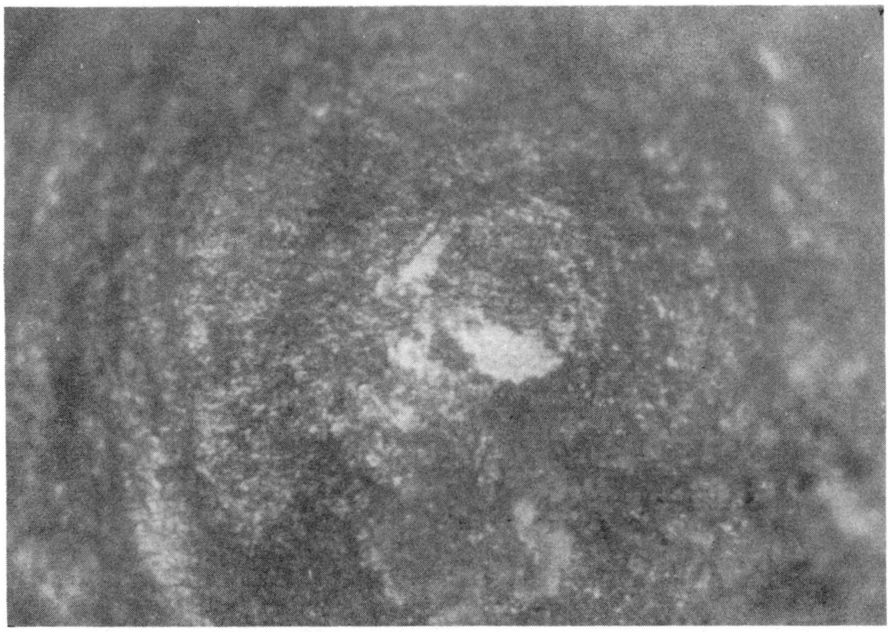

b [200x]

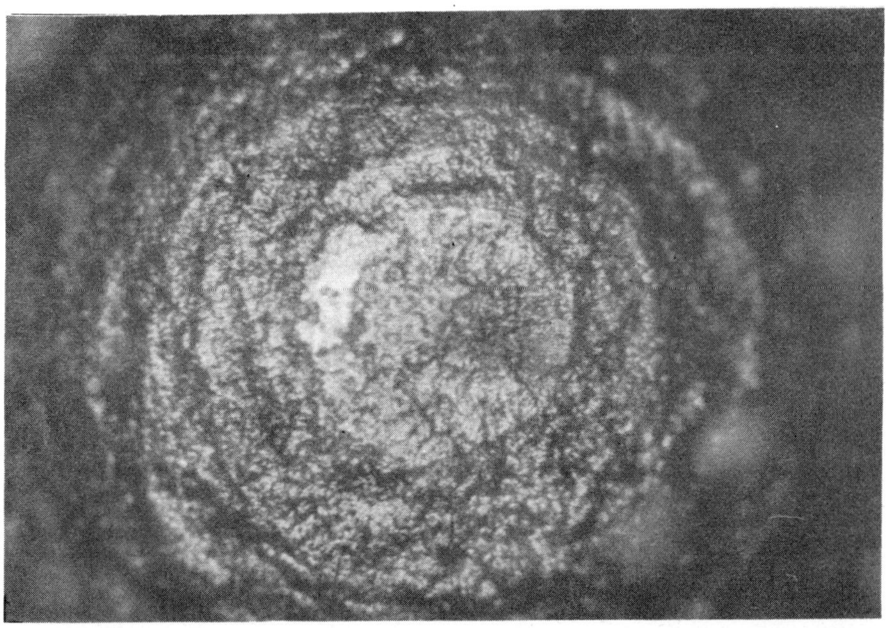

PLATE 13

circular striations around the perforation of a carnelian bead from Kumartepe [100x]

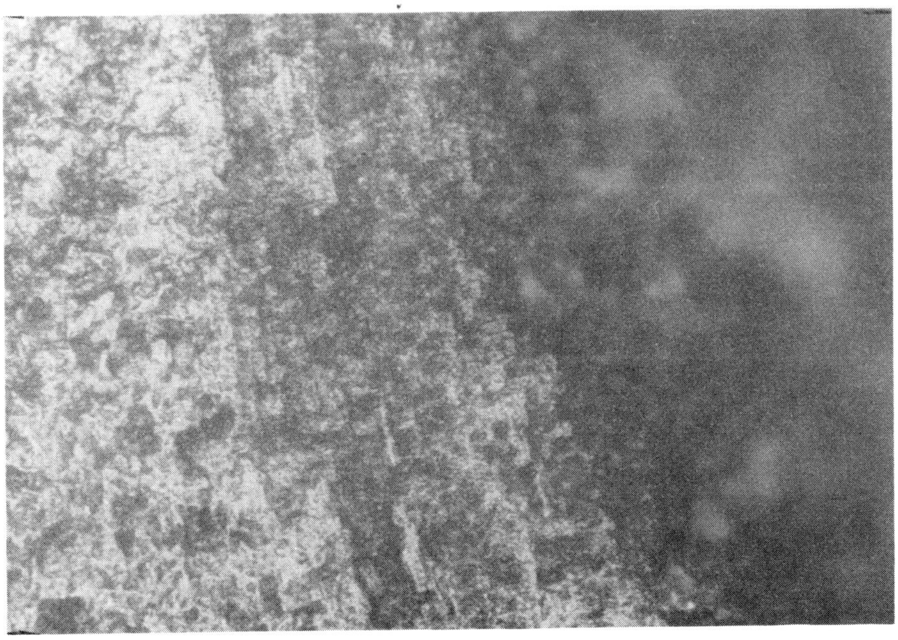

Figure 92

TOOL # 1 — **TYPE** endscraper

GRAINSIZE	coarse
TOPOGRAPHY	flat
TOPOGRAPHIC FEATURES	absent

EDGE ANGLE	70
LENGTH	20
THICKNESS	7
PROFILE	0.25
SHAPE	0.59

Find

EDGE WEAR

	VENTRAL macro	VENTRAL micro	DORSAL macro	DORSAL micro
GLOSS	absent		retouch	
FRACTURES	absent	absent	retouch	retouch
FRACTURE TYPE	absent	absent	retouch	retouch
ROUNDING	absent	absent	retouch	retouch

	VENTRAL	DORSAL
MICROTOPOGRAPHY	flat	retouch
DISTRIBUTION	continuous	retouch
DISTRIBUTION TYPE	edge only even	retouch
INVASIVENESS	edge only	retouch
LINEAR FEATURES	perpendicular	retouch
STRIATIONS	absent	retouch
POLISH DEV.	B	retouch

DATA TRANSFER

TOOL # 2 — **TYPE** drill bit POINT

GRAINSIZE	coarse
TOPOGRAPHY	flat
TOPOGRAPHIC FEATURES	absent

EDGE ANGLE	55
LENGTH	72
THICKNESS	3
PROFILE	2.83
SHAPE	0.35

Find

EDGE WEAR

	VENTRAL macro	VENTRAL micro	DORSAL macro	DORSAL micro
GLOSS	absent		absent	
FRACTURES	<5	>5	retouch	>5
FRACTURE TYPE	torsion	torsion	retouch	torsion
ROUNDING	light	heavy	light	heavy

	VENTRAL	DORSAL
MICROTOPOGRAPHY	flat	flat
DISTRIBUTION	absent	absent
DISTRIBUTION TYPE	absent	absent
INVASIVENESS	absent	absent
LINEAR FEATURES	absent	absent
STRIATIONS	absent	absent
POLISH DEV.	absent	absent

DATA TRANSFER

TOOL # 3 — **TYPE** piercer

GRAINSIZE	fine
TOPOGRAPHY	flat
TOPOGRAPHIC FEATURES	absent

EDGE ANGLE	59
LENGTH	10
THICKNESS	4
PROFILE	0
SHAPE	0.21

Find

EDGE WEAR

	VENTRAL macro	VENTRAL micro	DORSAL macro	DORSAL micro
GLOSS	absent		absent	
FRACTURES	absent	absent	retouch	retouch
FRACTURE TYPE	absent	absent	retouch	retouch
ROUNDING	absent	absent	retouch	retouch

	VENTRAL	DORSAL
MICROTOPOGRAPHY	flat	retouch
DISTRIBUTION	continuous	retouch
DISTRIBUTION TYPE	gapped	retouch
INVASIVENESS	<0.5	retouch
LINEAR FEATURES	absent	retouch
STRIATIONS	absent	retouch
POLISH DEV.	B+	retouch

DATA TRANSFER

TOOL # 4 — **TYPE** endscraper

GRAINSIZE	medium
TOPOGRAPHY	flat
TOPOGRAPHIC FEATURES	absent

EDGE ANGLE	70
LENGTH	39
THICKNESS	5.5
PROFILE	0.19
SHAPE	1.09

Find

EDGE WEAR

	VENTRAL macro	VENTRAL micro	DORSAL macro	DORSAL micro
GLOSS	absent		retouch	
FRACTURES	absent	<5	retouch	retouch
FRACTURE TYPE	absent	flakes	retouch	retouch
ROUNDING	light	heavy	retouch	retouch

	VENTRAL	DORSAL
MICROTOPOGRAPHY	flat	retouch
DISTRIBUTION	continuous	retouch
DISTRIBUTION TYPE	edge only even	retouch
INVASIVENESS	<0.5	retouch
LINEAR FEATURES	absent	retouch
STRIATIONS	absent	retouch
POLISH DEV.	C	retouch

DATA TRANSFER

TOOL # 5 — **TYPE** burin,facet

GRAINSIZE	medium
TOPOGRAPHY	undulating
TOPOGRAPHIC FEATURES	percussion ripples

EDGE ANGLE	77
LENGTH	10
THICKNESS	3.5
PROFILE	0.09
SHAPE	3

Find

EDGE WEAR

	VENTRAL macro	VENTRAL micro	DORSAL macro	DORSAL micro
GLOSS	absent		absent	
FRACTURES	absent	>5	absent	>5
FRACTURE TYPE	absent	flakes	absent	flakes&steps
ROUNDING	light	light	light	light

	VENTRAL	DORSAL
MICROTOPOGRAPHY	flat	flat
DISTRIBUTION	continuous	absent
DISTRIBUTION TYPE	edge only even	absent
INVASIVENESS	<0.5	absent
LINEAR FEATURES	absent	absent
STRIATIONS	absent	absent
POLISH DEV.	B+	absent

DATA TRANSFER

TOOL # 6 — **TYPE** azilian POINT

GRAINSIZE	medium
TOPOGRAPHY	undulating
TOPOGRAPHIC FEATURES	percussion ripples

EDGE ANGLE	70
LENGTH	96
THICKNESS	2
PROFILE	3.58
SHAPE	2.69

Find

EDGE WEAR

	VENTRAL macro	VENTRAL micro	DORSAL macro	DORSAL micro
GLOSS	absent		absent	
FRACTURES	<5	<5	<5	absent
FRACTURE TYPE	flute	flute	flute	absent
ROUNDING	absent	absent	absent	absent

	VENTRAL	DORSAL
MICROTOPOGRAPHY	flat	flat
DISTRIBUTION	absent	absent
DISTRIBUTION TYPE	absent	absent
INVASIVENESS	absent	absent
LINEAR FEATURES	absent	parallel
STRIATIONS	absent	absent
POLISH DEV.	absent	B+

DATA TRANSFER

Figure 92

TOOL # 7 TYPE endscraper

GRAINSIZE	coarse
TOPOGRAPHY	flat
TOPOGRAPHIC FEATURES	absent
EDGE ANGLE	60
LENGTH	22
THICKNESS	5
PROFILE	0.3
SHAPE	0.37

EDGE WEAR	VENTRAL macro	VENTRAL micro	DORSAL macro	DORSAL micro
GLOSS	absent		retouch	
FRACTURES	absent	absent	retouch	retouch
FRACTURE TYPE	absent	absent	retouch	retouch
ROUNDING	absent	absent	retouch	retouch

	VENTRAL	DORSAL
MICROTOPOGRAPHY	ridged	retouch
DISTRIBUTION	intermittent	retouch
DISTRIBUTION TYPE	edge only even	retouch
INVASIVENESS	edge only	retouch
LINEAR FEATURES	absent	retouch
STRIATIONS	absent	retouch
POLISH DEV.	B	retouch

Find · DATA TRANSFER

TOOL # 8 TYPE backed bladelet

GRAINSIZE	medium
TOPOGRAPHY	flat
TOPOGRAPHIC FEATURES	absent
EDGE ANGLE	23
LENGTH	31
THICKNESS	2
PROFILE	0.03
SHAPE	4.38

EDGE WEAR	VENTRAL macro	VENTRAL micro	DORSAL macro	DORSAL micro
GLOSS	absent		absent	
FRACTURES	<5	>5	<5	>5
FRACTURE TYPE	snaps	snaps	snaps	snaps
ROUNDING	absent	absent	absent	absent

	VENTRAL	DORSAL
MICROTOPOGRAPHY	flat	flat
DISTRIBUTION	absent	absent
DISTRIBUTION TYPE	absent	absent
INVASIVENESS	absent	absent
LINEAR FEATURES	absent	absent
STRIATIONS	absent	absent
POLISH DEV.	absent	absent

Find · DATA TRANSFER

TOOL # 9 TYPE blade

GRAINSIZE	fine
TOPOGRAPHY	ridged
TOPOGRAPHIC FEATURES	edge feathering
EDGE ANGLE	26
LENGTH	88
THICKNESS	5
PROFILE	0.02
SHAPE	5.18

EDGE WEAR	VENTRAL macro	VENTRAL micro	DORSAL macro	DORSAL micro
GLOSS	absent		absent	
FRACTURES	<5	>5	<5	>5
FRACTURE TYPE	snaps	flakes&snaps	snaps	flakes&steps
ROUNDING	absent	absent	absent	absent

	VENTRAL	DORSAL
MICROTOPOGRAPHY	ridged	flat
DISTRIBUTION	intermittent	continuous
DISTRIBUTION TYPE	edge only even	edge only even
INVASIVENESS	<0.5	<0.5
LINEAR FEATURES	absent	absent
STRIATIONS	absent	absent
POLISH DEV.	B	B+

Find · DATA TRANSFER

TOOL # 10 TYPE burin,facet, dihedral

GRAINSIZE	fine
TOPOGRAPHY	flat
TOPOGRAPHIC FEATURES	absent
EDGE ANGLE	90
LENGTH	9
THICKNESS	4.5
PROFILE	0
SHAPE	2.33

EDGE WEAR	VENTRAL macro	VENTRAL micro	DORSAL macro	DORSAL micro
GLOSS	absent		absent	
FRACTURES	absent	absent	absent	<5
FRACTURE TYPE	absent	absent	absent	flakes
ROUNDING	absent	light	absent	light

	VENTRAL	DORSAL
MICROTOPOGRAPHY	flat	flat
DISTRIBUTION	absent	continuous
DISTRIBUTION TYPE	absent	edge only even
INVASIVENESS	absent	edge only
LINEAR FEATURES	absent	absent
STRIATIONS	absent	absent
POLISH DEV.	absent	A+

Find · DATA TRANSFER

Figure 93

MACRO FRACTURE TYPE

VENTRAL: absent | DORSAL: retouch

	VENTRAL	DORSAL
FLAKES & ABSENT	***NON-DIAGNOSTIC***	NON DIAGNOSTIC
SNAPS	CUTTING	CUTTING
STEPS	PERCUSSIVE MOTION	PERCUSSIVE MOTION
FLAKES AND STEPS	if angle ≥50 then PERCUSSIVE	if angle ≥50 then PERCUSSIVE
	if angle < 50 then SCRAPING MEDIUM	if angle < 50 then SCRAPING MEDIUM
FLAKES AND SNAPS	CUTTING MEDIUM	CUTTING MEDIUM
RETOUCH	RETOUCH	***RETOUCH***
OTHER COMBINATIONS	EDGE DAMAGE	EDGE DAMAGE

edge angle = 42

MACRO ROUNDING

VENTRAL: absent | DORSAL: retouch

	VENTRAL	DORSAL
LIGHT if angle ≤45	SOFTor SHORT TIME [MEDIUM]	SOFTor SHORT TIME [MEDIUM]
if angle >45	MEDIUM	MEDIUM
HEAVY if angle ≤45	SOFT	SOFT
if angle >45	HARD or LONG TIME [MEDIUM]	HARD or LONG TIME [MEDIUM]
ABSENT if angle ≤30	SOFT	SOFT
if angle 30-60	***SOFT to MEDIUM***	SOFT to MEDIUM
if angle >60	NON DIAGNOSTIC	NON DIAGNOSTIC
RETOUCH	RETOUCH	***RETOUCH***

edge angle = 42

MICRO FRACTURES

VENTRAL: <3 | DORSAL: retouch

	VENTRAL	DORSAL
ABSENT if angle <60	SOFT MATERIAL	SOFT MATERIAL
if angle ≥60	NON DIAGNOSTIC	NON DIAGNOSTIC
<5 if angle <30	SOFT MATERIAL	SOFT MATERIAL
if angle 30-60	***SOFT to MEDIUM***	SOFT to MEDIUM
if angle >60	HARD MATERIAL	HARD MATERIAL
>5 if angle <30	SOFT to MEDIUM	SOFT to MEDIUM
if angle 30-60	MEDIUM to HARD	MEDIUM to HARD
if angle >60	HARD MATERIAL	HARD MATERIAL
RETOUCH	RETOUCH	***RETOUCH***

EDGE ANGLE= 42

MICRO FRACTURE TYPE

VENTRAL: flakes | DORSAL: retouch

	VENTRAL	DORSAL
FLAKES & ABSENT	***NON-DIAGNOSTIC***	NON DIAGNOSTIC
SNAPS	CUTTING	CUTTING
STEPS	PERCUSSIVE on MEDIUM	PERCUSSIVE on MEDIUM
FLAKES AND STEPS	if angle > 50 SCRAPING HARD	SCRAPING HARD
	if angle < 50 then SCRAPING MEDIUM	if angle < 50 then SCRAPING MEDIUM
FLAKES AND SNAPS	CUTTING MEDIUM	CUTTING MEDIUM
RETOUCH	RETOUCH	***RETOUCH***
OTHER COMBINATIONS	EDGE DAMAGE	EDGE DAMAGE

edge angle = 42

MICRO ROUNDING

VENTRAL: absent | DORSAL: retouch

	VENTRAL	DORSAL
LIGHT if angle ≤45	SOFTor SHORT TIME [MEDIUM]	SOFTor SHORT TIME [MEDIUM]
if angle >45	MEDIUM	MEDIUM
if angle >60	HARD MATERIAL	HARD MATERIAL
HEAVY if angle ≤60	USED FOR A LONG TIME	USED FOR A LONG TIME
if angle >60	HARD or LONG TIME [MEDIUM]	HARD or LONG TIME [MEDIUM]
ABSENT if angle ≤30	SOFT	SOFT
if angle 30-60	***SOFT to MEDIUM***	SOFT to MEDIUM
if angle >60	NON DIAGNOSTIC	NON DIAGNOSTIC
RETOUCH	RETOUCH	***RETOUCH***

edge angle = 42

MICROTOPOGRAPHY

VENTRAL: flat | DORSAL: retouch

	VENTRAL	DORSAL
FLAT	***NO EFFECT***	NO EFFECT
UNDULATING	may cause, INT. DIST.	may cause INT. DIST.
RIDGED	may cause LINEAR or INT. DIST	may cause LINEAR or INT. DIST
RETOUCH	RETOUCH	***RETOUCH***

POLISH DISTRIBUTION

VENTRAL: continuous | DORSAL: retouch

	VENTRAL	DORSAL
CONTINUOUS	***NON DIAGNOSTIC***	NON DIAGNOSTIC
INTERMITTENT length ≤10	[see TOP] HARD	[see TOP] HARD
length >10	see TOP or high contact [SOFT]	see TOP or high contact [SOFT]
ABSENT	NO POLISH	NO POLISH
RETOUCH	RETOUCH	***RETOUCH***

length = 28

DISTRIBUTION TYPE

VENTRAL: edge only even | DORSAL: retouch

	VENTRAL	DORSAL
AWAY FROM EDGE	GROOVING MEDIUM TO HARD	GROOVING MEDIUM to HARD
GAPPED	GROOVING OR CUTTING HARD	GROOVING or CUTTING HARD
EDGE ONLY EVEN	***NON DIAGNOSTIC***	NON DIAGNOSTIC
EDGE ONLY ASYMMETRIC	GROOVING/CUTTING	GROOVING/CUTTING
DIFFERENTIAL	see TOP or 2 MATERIALS	see TOP or 2 MATERIALS
ABSENT	NO POLISH	NO POLISH
RETOUCH	RETOUCH	***RETOUCH***

Figure 93

TOOL # 33 TYPE endscraper

GRAINSIZE	fine
TOPOGRAPHY	flat
TOPOGRAPHIC FEATURES	absent

EDGE ANGLE	42	⊠
LENGTH	28	⊠
THICKNESS	6	⊠
PROFILE	0.56	⊠
SHAPE	0.53	⊠

EDGE WEAR

	VENTRAL macro	VENTRAL micro	DORSAL macro	DORSAL micro
GLOSS	⊠ absent		retouch	
FRACTURES	⊠ absent	<5	retouch	retouch ⊠
FRACTURE TYPE	⊠ absent	flakes	retouch	retouch ⊠
ROUNDING	⊠ absent	absent	retouch	retouch ⊠

MICROTOPOGRAPHY	⊠ flat	retouch
DISTRIBUTION	⊠ continuous	retouch
DISTRIBUTION TYPE	⊠ edge only even	retouch
INVASIVENESS	⊠ <0.5	retouch
LINEAR FEATURES	perpendicular	retouch
STRIATIONS	⊠ perpendicular	retouch
POLISH DEV.	⊠ C	C

EDGE ANGLE 42

<30	CUTTING SOFT
30-55	***CUTTING OR SCRAPING MEDIUM***
55-80	MEDIUM TO HARD MATERIAL
>80	SCRAPING HARD

EDGE LENGTH 28

≤5	GROOVING HARD
>5 <10	SCRAPING/GROOVING HARD
≥10 <35	***SCRAPING***
≥35	CUTTING

THICKNESS 6

1-4	CUTTING
>4	***SCRAPING***

PROFILE 0.56

< 0	SCRAPING MEDIUM OR HARD
≥0 ≤0.1	CUTTING
>0.1 <1	***SCRAPING***
≥1	POINT

SHAPE 0.53

≤2	***SCRAPING***
>2	CUTTING
>2 & burin	SCRAPING/ GROOVING HARD
IF POINT	NON DIAGNOSTIC

endscraper

GLOSS

VENTRAL absent DORSAL retouch

	VENTRAL	DORSAL
ABSENT	***NON DIAGNOSTIC***	NON DIAGNOSTIC
RETOUCH	RETOUCH	***RETOUCH***
PRESENT angle ≥60	HARD MATERIAL	HARD MATERIAL
angle >40 <60	MEDIUM MATERIAL	MEDIUM MATERIAL
angle ≤40	SOFT for a LONG TIME	SOFT for a LONG TIME

edge angle= 42

MACRO FRACTURES

VENTRAL absent DORSAL retouch

	VENTRAL	DORSAL
ABSENT angle ≤30	SOFT MATERIAL	SOFT MATERIAL
angle >30 <60	***MEDIUM MATERIAL***	MEDIUM MATERIAL
angle ≥60	HARD MATERIAL	HARD MATERIAL
<5	SOFT TO MEDIUM MATERIAL	SOFT TO MEDIUM MATERIAL
>5 angle <30	SOFT MATERIAL	SOFT MATERIAL
angle ≥30 <50	MEDIUM MATERIAL	MEDIUM MATERIAL
angle ≥50	MEDIUM to HARD	MEDIUM TO HARD
RETOUCH	RETOUCH	***RETOUCH***

EDGE ANGLE= 42

Figure 93

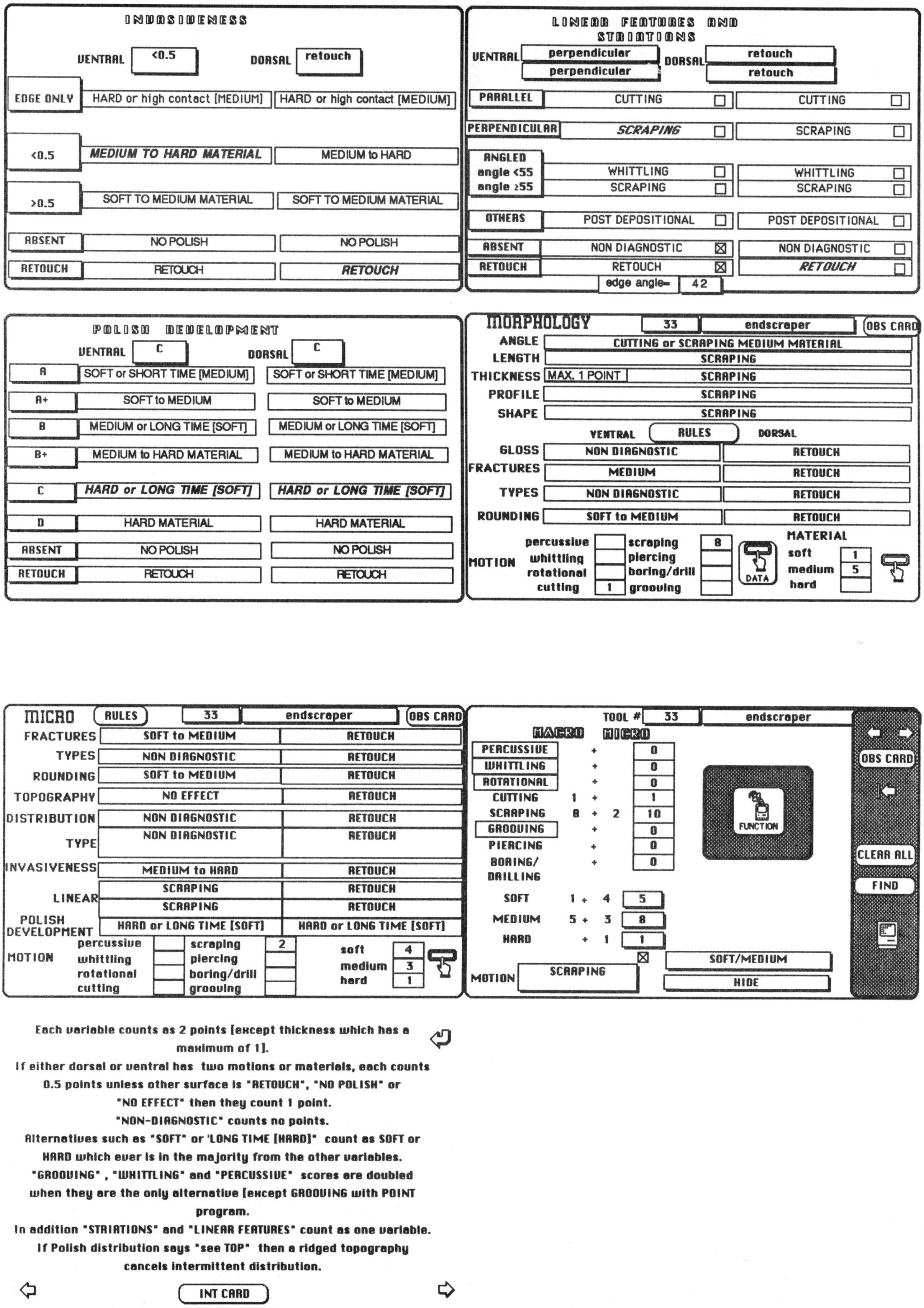
INVASIVENESS
VENTRAL <0.5 DORSAL retouch
EDGE ONLY HARD or high contact [MEDIUM] HARD or high contact [MEDIUM]
<0.5 MEDIUM TO HARD MATERIAL MEDIUM to HARD
>0.5 SOFT TO MEDIUM MATERIAL SOFT TO MEDIUM MATERIAL
ABSENT NO POLISH NO POLISH
RETOUCH RETOUCH RETOUCH
LINEAR FEATURES AND STRIATIONS
VENTRAL perpendicular perpendicular DORSAL retouch retouch
PARALLEL CUTTING CUTTING
PERPENDICULAR SCRAPING SCRAPING
ANGLED angle <55 angle ≥55 WHITTLING SCRAPING WHITTLING SCRAPING
OTHERS POST DEPOSITIONAL POST DEPOSITIONAL
ABSENT NON DIAGNOSTIC NON DIAGNOSTIC
RETOUCH RETOUCH RETOUCH
edge angle= 42
POLISH DEVELOPMENT
VENTRAL C DORSAL C
A SOFT or SHORT TIME [MEDIUM] SOFT or SHORT TIME [MEDIUM]
A+ SOFT to MEDIUM SOFT to MEDIUM
B MEDIUM or LONG TIME [SOFT] MEDIUM or LONG TIME [SOFT]
B+ MEDIUM to HARD MATERIAL MEDIUM to HARD MATERIAL
C HARD or LONG TIME [SOFT] HARD or LONG TIME [SOFT]
D HARD MATERIAL HARD MATERIAL
ABSENT NO POLISH NO POLISH
RETOUCH RETOUCH RETOUCH
MORPHOLOGY 33 endscraper OBS CARD
ANGLE CUTTING or SCRAPING MEDIUM MATERIAL
LENGTH SCRAPING
THICKNESS MAX. 1 POINT SCRAPING
PROFILE SCRAPING
SHAPE SCRAPING
VENTRAL RULES DORSAL
GLOSS NON DIAGNOSTIC RETOUCH
FRACTURES MEDIUM RETOUCH
TYPES NON DIAGNOSTIC RETOUCH
ROUNDING SOFT to MEDIUM RETOUCH
MOTION percussive whittling rotational cutting 1 scraping 8 piercing boring/drill grooving
DATA
MATERIAL soft 1 medium 5 hard
MICRO RULES 33 endscraper OBS CARD
FRACTURES SOFT to MEDIUM RETOUCH
TYPES NON DIAGNOSTIC RETOUCH
ROUNDING SOFT to MEDIUM RETOUCH
TOPOGRAPHY NO EFFECT RETOUCH
DISTRIBUTION NON DIAGNOSTIC RETOUCH
TYPE NON DIAGNOSTIC RETOUCH
INVASIVENESS MEDIUM to HARD RETOUCH
LINEAR SCRAPING RETOUCH SCRAPING RETOUCH
POLISH DEVELOPMENT HARD or LONG TIME [SOFT] HARD or LONG TIME [SOFT]
MOTION percussive whittling rotational cutting scraping 2 piercing boring/drill grooving
soft 4 medium 3 hard 1
TOOL # 33 endscraper
MACRO MICRO
PERCUSSIVE + 0
WHITTLING + 0
ROTATIONAL + 0
CUTTING 1 + 1
SCRAPING 8 + 2 10
GROOVING + 0
PIERCING + 0
BORING/ DRILLING + 0
SOFT 1 + 4 5
MEDIUM 5 + 3 8
HARD + 1 1
SOFT/MEDIUM
HIDE
MOTION SCRAPING
FUNCTION
OBS CARD
CLEAR ALL
FIND
Each variable counts as 2 points [except thickness which has a maximum of 1].
If either dorsal or ventral has two motions or materials, each counts 0.5 points unless other surface is "RETOUCH", "NO POLISH" or "NO EFFECT" then they count 1 point.
"NON-DIAGNOSTIC" counts no points.
Alternatives such as "SOFT" or 'LONG TIME [HARD]" count as SOFT or HARD which ever is in the majority from the other variables.
"GROOVING", "WHITTLING" and "PERCUSSIVE" scores are doubled when they are the only alternative [except GROOVING with POINT program.
In addition "STRIATIONS" and "LINEAR FEATURES" count as one variable.
If Polish distribution says "see TOP" then a ridged topography cancels intermittent distribution.
INT CARD

Figure 94

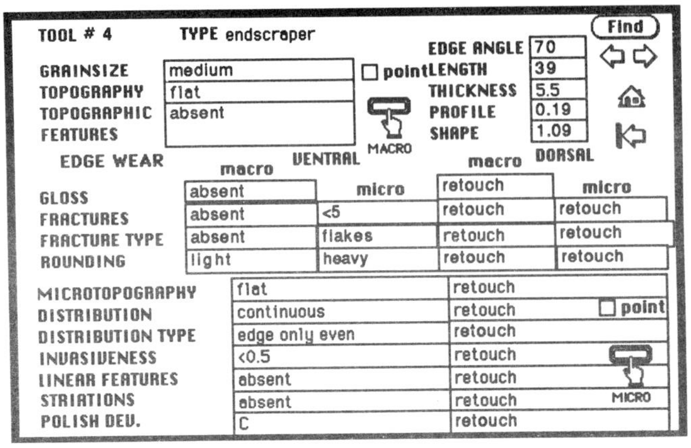

For the VERY FAST program the program for each variable has been combined so that on the DATA CARD only two buttons have to be pressed the MACRO button and the MICRO button.

(For points the two point buttons are pressed. ☐ point)

The program then puts all the indications into the two DIAGNOSTIC CARDS. The indications are counted according to the rules and entered into the INTERPRETATION CARD. The FUNCTION program (FUNCTION) is activated and the interpretation presented in the boxes. The data and interpretation of Tool 4 is presented in this example.

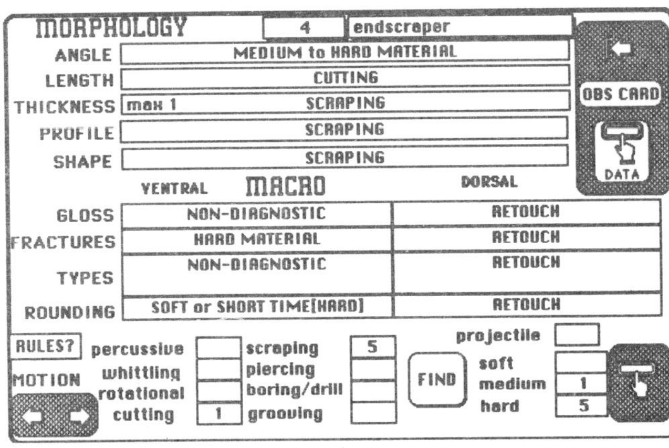

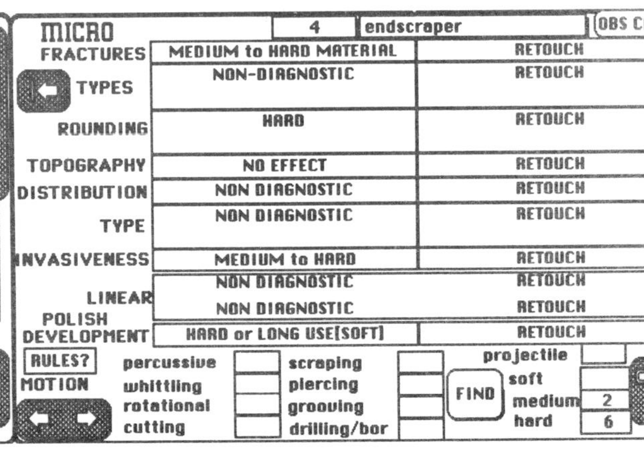

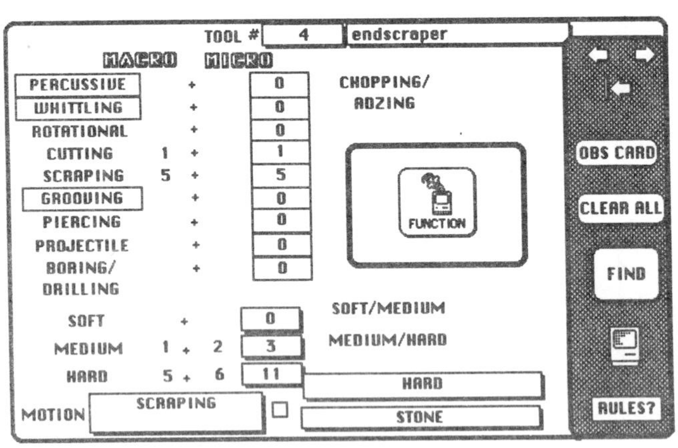

THE FUNCTION OF OTHER BUTTONS

RULES? Accesses card which contains the rules.

OBS CARD Accesses data card with tool observations.

DATA Accesses data base.

CLEAR ALL Clears all fields ready for next analysis.

FIND Used to find any "string" in stack.

Closes Hypercard and returns to Macintosh operation.

FAST INTERPRETATIONS

#	MOTION	HARDNESS	MATERIAL	CORRECT MATERIAL
1	scraping	medium	wood	correct
2	boring/drilling	medium	unknown	SHELL
3	grooving	medium/hard	antler	correct
4	scraping	hard	stone	correct
5	scraping	hard	bone	correct
6	projectile	medium	insuf. data	unknown
7	---	-------	------	unused
8	cutting	soft	meat	correct
9	cutting/sawing	soft/medium	fish	correct
10	scraping	medium/hard	antler	WOOD
31	grooving	medium/hard	s.antler [horn]	correct
32	scraping	medium/hard	antler	correct
33	scraping	soft/medium	hide	correct
34	grooving	medium/hard	s.antler[horn]	correct
35	cutting	soft/medium	soft wood	correct
36	boring/drilling	medium	s.antler [horn]	correct
37	cutting	soft/medium	soft wood	correct
38	piercing	soft	insufficient data	BARK
39	scraping	hard	bone	correct
40	piercing	soft/medium	hide	correct
41	scraping	soft medium	soft wood	correct
42	grooving	hard	shell	correct
43	chop/adzing	soft/medium	wood	correct
44	cutting	soft medium	unknown	CORTEX
45	grooving	medium/hard	s. antler[horn]	correct
46	cutting/sawing	soft/medium	fish	correct
47	scraping	medium	wood	correct
48	scraping	hard	bone	correct
49	scraping	medium/hard	antler	correct
50	whittling	medium	wood	correct

BIBLIOGRAPHY

Akoshima,K. 1981. An experimental study of microflaking - report of Tohoku University Microflaking Research Team. *Kokogaku Zasshi* (Journal of the Archaeological Society of Nippon) vol.66 (in Japanese).

Akoshima, Kaoru. 1987. Microflaking Quantification. In *The Human Uses of Flint and Chert:* papers presented at the Fourth International Flint Symposium. eds. Sieveking, G. de G. and Newcomer, M.H. Cambridge University Press. Cambridge. 71-80.

Anderson-Gerfaud, P. 1981. *Contribution Méthodologique a l'analyse des microtraces d'utilisation sur les outils préhistoriques.* Thèse de 3e cycle. Université de Bordeaux.

Barton, R. and Bergman, C. 1982. Hunters at Hengisbury; some evidence from experimental archaeology. *World Archaeology* 14(2);237-48

Binford, L.R. and Binford, S. 1966. A Preliminary Analysis of Functional Variability in the Mousterian of Levallois facies. *American Anthropologist* v. 68 no. 2, 238-295.

Brink, J.W. 1978. *Experimental study of microwear formation on endscrapers.* M.A. Thesis. University of Alberta.

Bamforth, D.B. 1987. Investigating Microwear polishes with Blind Tests: The Institute results in Context. *Journal of Archaeological Science.*

Bordes, F. 1961. *Typologie du Paléolithique Ancien et Moyen.* Bordeaux.

Bordes, F. 1972. *A Tale of Two Caves.* Bordeaux.

Broadbent, N. 1979. *Coastal Settlement and Settlement Stability.* Borgtrans. Uppsala.

Büller, H. 1983. Methodological problems in microwear analysis of tools selected from the Natufian sites of El Wad and Ain Mallaha. In Cauvin, M-C, ed. *Traces d'Utilisation sur les outils Néolithique du Proche Orient,* 107-127.

Cahen, D., Keeley, L.H. and Van Noten, F.L. 1979. Stone tools, tool kits and human behavior in Prehistory. *Current Anthropology* 20 (4),661-683.

Calley, S. and Grace, R. 1988. Drill bits from Kumartepe, Turkey. In *Industries l ithiques: tracéologie et technologie.* B.A.R. International Series 411(1), 69-81.

Cantwell, A, 1979. The functional analysis of scrapers: Problems, new techniques and cautions. *Lithic Technology* 8 (1), 5-9.

Chevalier, J., Inizan, M-L. and Tixier, J. 1982. Une technique de perforation par percussion de perles en cornaline (Larsa, Iraq). *Paléorient* 8(2), 55-65

Cook, J. and Dumont, J. 1987. The Development and application of microwear analysis since 1964. In *The Human Uses of Flint and Chert:* papers presented at the Fourth International Flint Symposium. eds. Sieveking, G. de G. and Newcomer, M.H. Cambridge University Press. Cambridge. 53-61

Darling Jr., E.M. and Joseph, R.D. 1968. Pattern recognition from satellite altitudes. *IEEE Transactions. Computers.* C-20 (9), 1032-44.

Dumont, J. 1982. The quantification of microwear traces: a new use for interferometry. *World Archaeology* 14, 206-217.

Dibble, H.L. and Bernard, M.C. 1980 A comparative study of basic edge angle measurement techniques. *American Antiquity* 45(4), 837-865.

Driskell, B. 1986. *The chipped stone production/use cycle.* B.A.R. International series no. 305.

Fischer,A., Hansen,P.V., and Rasmussen,P. 1984. Macro and micro wear traces on lithic projectile points. *Journal of Danish Archaeology* 3:19-46

Frison, G.C. 1968. A functional analysis of certain chipped stone tools. *American Antiquity* 33(2), 149-155

Gendel, P.A. and Pirnay, L. 1982. Microwear Analysis of Experimental Stone Tools: Further Test Results. *Studia Prehistorica Belgica.* 2, 251-65.

Gould, R., Koster, D. and Sontz, A. 1971. The lithic assemblage of the Western Desert aborigines of Australia. *American Antiquity* 36 (2), 149-169

Grace, R. 1981. *Attribute Analysis as a Method of Studying Functional Variability in Palaeolithic Assemblages.* B.A. dissertation. University of London.

Grace, R., Graham, I.D.G. and Newcomer, M.H. 1985. The Quantification of Microwear Polishes.*World Archaeology.* 17(1), 112-120.

Grace, R., Ataman, K., Fabregas, R., and Haggren, C.M.B. 1988. A multivariate approach to the functional analysis of stone tools. In *Industries lithiques: tracéologie et technologie.* B.A.R. International Series 411(2),217-230.

Hall, E.I., Kruger, R.P., Dwyer, S.J., Hall, D.L., McLaren, L.W. and Hardwick, G.S. 1971. A survey of preprocessing and feature extraction techniques for radiographic images. *IEEE Transactions.Computers*. C-20 no. 9,1032-44.

Haralick, R.M., Shanmugan, K. and Dinstein, L. 1973. Texture features for image classification. *IEEE Transactions. Systems, Man and Cybernetics* SMC-3, 610-621

Hurcombe, L. 1988. Some criticisms and suggestions in response to Newcomer *et al.* *Journal of Archaeological Science* .

Kamminga, J. 1982 Over The Edge: functional analysis of Australian stone tools. *Occasional Papers in Anthropology* no.12 University of Queensland.

Keeley, L.H. and Newcomer, M.H. 1977. Microwear Analysis of Experimental Flint Tools: a Test Case. *Journal of Archaeological Science*. 4(1), 29-62.

Keeley, L.H. 1977. The function of Palaeolithic flint tools. *Scientific American* 237(4), 108-26.

Keeley, L.H. 1980. *Experimental Determination of Stone Tool Uses: a Microwear Analysis*. University of Chicago Press. Chicago.

Keeley, L.H. 1982. Hafting and Re-tooling; Effect on the archaeological record. *American Antiquity* 47, 798-805.

Knutsson, K. 1985. SEM-analysis of Wear features on experimental Quartz tools. *Early Man News* 9/10/11. Tübingen.

Knutsson, K., Dahlquist, B. and Knutsson, H. 1988. Patterns of tool use; the microwear analysis of the quartz and flint assemblage from the Bjurselet site, Västerbotten, Northern Sweden. In *Industries lithiques: tracéologie et technologie*. B.A.R. International Series 411(1), 253-294.

Levi Sala, I. 1986. A Word of Caution: Use Wear and Post-depositional Surface Modification. *Journal of Archaeological Science* 13,

Loy, T. 1983. Prehistoric blood residues: Detection on tool surfaces and identification of species of origin. *Science* 220.

Mandelbrot, B.B. 1982. *The Fractal Geometry of Nature*. W.H. Freeman. New York.

Mansur, M.E. 1982. Microwear analysis of natural and use striations: New clues to the mechanism of striation formation. *Studia Prehistorica Belgica*. 2, 213-233

Moss, E. and Newcomer, M.H. 1982. Reconstruction of tool use at Pincevent: Microwear and experiments. Tailler! pour quoi faire: Préhistoire et technologie lithique 2. Recent progress in microwear studies. *Studia Prehistorica Belgica* 2 D. Cahen ed. Terveuren: Musee royale de l'Afrique central 289-312.

Moss, E.H. 1983. *The Functional Analysis of Flint Implements*. British Archaeological Reports, International series no. 177.

Moss, E. 1987 A review of 'Investigating Microwear Polishes with Blind Tests'. *Journal of Archaeological Science* 13.

Movius, H.L.Jr., and Brooks, A.S. 1971. The analysis of certain major classes of Upper Palaeolithic tools: Aurignacian scrapers. *Proceedings of the Prehistoric Society* 37.

Newcomer, M.H., Grace, R. and Unger-Hamilton, R. 1986. Investigating microwear polishes with blind tests. *Journal of Archaeological Science.* 13, 203-217.

Odell, G,.and Odell-Vereecken, F. 1981. Verifying the reliability of lithic use wear assessment by 'blind tests': the low power approach. *Journal of Field Archaeology* 7(1), 87-120.

Olaussen, D.S. 1978. Starting from scratch: The history of edge wear research from 1838-1978. *Lithic Technology* 9.

Panagopoulou, H. 1985. Form versus Function: lithic use-wear analysis and its application to a class of Levantine Mousterian tools. Ph. D. Thesis. Columbia University.

Pratt, W.K. 1978. *Digital Image Processing*. New York. Wiley.

Rees,D., Wilkinson, G.G., Orton, C.R. and Grace,R. 1988. Fractal analysis of digital images of flint microwear. Presented to the C.A.A. in Birmingham.

Roodenberg, J.J., Wilkinson, J.J. and Bayri-Baykan, S. 1984. Surveys and soundings at Kumartepe; an interim report. *Anatolica* 11, 1-16.

Schutt, J.A. 1982. Comparative analysis of wear patterns from experimental lithic flake tools: The re-examination of current concepts in tool utilisation. M.A. thesis . University of New Mexico. Albuquerque, New Mexico.

Semenov, S.A. 1964. *Prehistoric Technology*. Adams and Dart. London

Tomenchuk, J. 1983. Predicting the past: examples from the use-wear study of selected chipped stones from two Epi-Palaeolithic occupations in Israel. In (M-C, Cauvin, Ed.)*Traces d'Utilisation sur les Outils Néolithiques du Proche Orient. Travaux de la Maison de l'Orient, Lyon.* 5, 57-76.

Tringham, R., Cooper, G., Odell, G., Voytek, B., and Whitman, A. 1974. Experimentation in the formation of edge damage: A new approach to lithic analysis. *Journal of Field Archaeology* 1, 171-196.

Unger-Hamilton, R., Grace, R., Miller, R. and Bergman, C. 1987. Drill bits from Abu Salabikh, Iraq. *(Table-Rond C.N.R.S. Manches et Emmanchements préhistoriques)* Lyon, Maison de l'Orient.

Unrath, G., Owen, L.R. , van Gijn, A., Moss, E.H., Plisson H. and Vaughan, P. 1986. An Evaluation of Use-wear Studies: a Multi-analyst Approach. *Early Man News* v.9/10/11,Part 1,117-175.

Vaughan, P. 1981. *Lithic Experimentation and the Functional Analysis of a Lower Magdelanian Stone Tool Assemblage.* Ph.D. dissertation. University of Pennsylvania.

Vaughan, P. 1985. *Use wear analysis of flaked stone tools.* University of Arizona Press. Tucson

White, J.P. 1969. Typologies for some prehistoric flaked stone artifacts in the Australian New Guinea Highlands. *Archaeology and Anthropology in Oceania* 4, 18-46.

White, J.P. and Thomas, D.H. 1972. What mean these stones ? Ethno-taxonomic models and archaeological interpretation in the New Guinea Highlands. In *Methods in Archaeology* ed. D.L. Clarke. Methuen. London. 275-308.

Wilmsen, E. 1968. Functional analysis of flaked stone artifacts. *American Antiquity* 22, 156-61.

INDEX

ACKNOWLEDGEMENTS

I would like to thank my colleagues at the Institute of Archaeology for the many hours of discussion about microwear analysis and for allowing me to look at the results of their experiments in order to augment my own.

For the Blind test I would like to thank Mark Newcomer and Chris Bergman for providing the tools and using them in their various and sometimes strange ways, and Kathy Ataman, Ramon Fabregas and Charlotte Haggren for taking part.

Most of the image processing techniques were developed in collaboration with Ian Graham.

The idea for an expert system computer program was developed with Zoe Tomlinson.

I would like to thank Jacob Roodenberg for his hospitality and invaluable advice during my research in Turkey, and I gratefully acknowledge the provision of funds for that research from the Dutch Institute of Archaeology, Istanbul and the Gordon Childe Fund.

I would like to acknowledge the help and encouragement of my supervisor Mark Newcomer, and in particular Roy Hodson whose interest in this research has been above and beyond the call of duty.

This monograph is dedicated to the memory of Frank Herbert who provided inspiration.

Last, and therfore foremost, I thank Charlotte for spiritual guidance.

Hello Mum & Fred!

GALILEO: ... I do not hope to prove I have been right up to now - but to find out if I have been. I say - abandon hope, all you who enter into an observation. Perhaps they're vapours. Perhaps they're spots. But before we say they are spots, let us say they are fishes tails. We'll question everything, everything all over again. And we won't run at it in great big boots, we'll go at a snails pace. And what we find today, we'll strike from the record tomorrow. And only when we find it once more will we write it in. And when we find something we want to find, we'll look at it with fierce suspicion. So - we will start our observation of the sun with the determination to prove the earth STANDS STILL! And only when we fail, when we are beaten and licking our wounds, shattered and depressed, will we ask - were we right, does the earth go around the sun ? (*he winks*) But if all possibilities but that one vanish into thin air, then - no mercy to they who did not observe but still shout their mouths off. Take the cloth off the telescope ! Point it at the sun !

The Life of Galileo .... by Bertolt Brecht